Electric Drive

Electric Drives

D.P. Kothari

Director Research, Gaikwad-Patil Group of Institutions, Nagpur
Former Director Research, MVSR Engineering College, Hyderabad
Former Director General, JBIET, Hyderabad
Former Director General, Raisoni Group of Institutions, Nagpur
Emeritus Director General, Vindhya Institute of Technology and Science
(VITS), Indore
Former Vice Chancellor, Vellore Institute of Technology (VIT), Vellore
Former Director In-Charge, Indian Institute of Technology (IIT) Delhi
Former Principal, Visvesvaraya National Institute of Technology (VNIT)
Nagpur

Rakesh Singh Lodhi

Assistant Professor
Department of Electrical & Electronics Engineering
Oriental University
Indore

I.K. International Publishing House Pvt. Ltd.

NEW DELHI

Published by
I.K. International Publishing House Pvt. Ltd.
S-25, Green Park Extension
Uphaar Cinema Market
New Delhi–110 016 (India)
E-mail:info@ikinternational.com
Website: www.ikbooks.com

ISBN: 978-93-84588-12-0

Published by Krishan Makhijani for I.K. International Publishing House Pvt. Ltd., S-25, Green Park Extension, Uphaar Cinema Market, New Delhi–110 016 and Printed by Rekha Printers Pvt. Ltd., Okhla Industrial Area, Phase II, New Delhi–110 020.

Preface

The present book is meant for undergraduate and postgraduate students in electrical, electrical & electronics, power electronics and automation courses of all engineering colleges (B.E./B.Tech./M.E./M.Tech./Ph.D.). It presents a unique self-study material on electrical drives, solid-state drives, industrial drives and power semiconductor drives. Solved and unsolved problems, multiple-choice questions of different patterns have been given to the students for practice. Chapter 1 describes the general introduction about drives. Chapter 2 gives information related to components, applications and factors affecting the selection of electrical drives. The remaining chapters provide all the required information about the electrical drives. A large number of solved and unsolved problems with answers make this book suitable for undergraduate and postgraduate courses in electrical drives. Practising engineers and those appearing for engineering services/ GATE examinations IES/IAS will also find this book useful.

Suggestions for further improvement of the book will be appreciated.

D. P. Kothari
Rakesh Singh Lodhi

Acknowledgements

I am also grateful to the authorities of GPGI, Dr. Mohan Gaikwad, Chairman and Col. Rahul Sharma, Director General for their constant encouragement.

I am thankful to Ms. Pooja Badole for her typing some portion of the manuscript.

D. P. Kothari

I appreciate the patience, understanding and support of my father, Mr. Dhiraj Singh Lodhi. I appreciate the knowledge and supporting material provided by Dr. H. K. Verma and Dr. Shailendra Sharma, Professors in Electrical Engineering Department, SGSITS, Indore.

I want to express my sincere thanks to Mr. Sunil Senani, Chairman of Vindhya Group of institutions for his co-operation, and motivation for the book.

I am grateful to Prof. Rahul Agrawal, Head of the Department, Prof. A. J. Siddiqui, Executive Director & Dr. R. S. Tare, Former Principal VITS, Indore for extending me all the possible facilities to carry out the work.

I am thankful to Dr. K . L . Thakral, Chancellor Oriental University, Indore for providing me resources for motivation and inspiration for this book.

I am thankful to Mrs. Neha Maithil Kothari, for her contribution in reviewing and editing the manuscript of the book. I am thankful to all my colleagues of Electrical & Electronics Department for their cooperation. I am also thankful to all VITS and Oriental University staff Indore for helping me in completing this work.

Rakesh Singh Lodhi

Contents

The Authors

D.P. Kothari is Director–Research, Gaikwad-Patil Group of Institutions, Nagpur. He obtained his BE (Electrical) in 1967, ME (Power Systems) in 1969 and PhD in 1975 from the Birla Institute of Technology and Science (BITS, Pilani), Rajasthan. From 1969 to 1977, he was involved in teaching and development of several courses at BITS, Pilani. Dr Kothari has also served as Director–Research, MVSR Engineering College, Hyderabad; Director General, Raisoni Group of Institutions, Nagpur; Emeritus Director General, Vindhya Institute of Technology and Science (VITS), Indore; and Vice Chancellor, Vellore Institute of Technology (VIT), Vellore. He has also served as Director In-charge and Deputy Director (Administration) as well as Head, Centre for Energy Studies at Indian Institute of Technology Delhi; and as Principal, Visvesvaryaya National Institute of Technology, Nagpur. He was Visiting Professor at the Royal Melbourne Institute of Technology, Melbourne, Australia, during 1982–83 and 1989 for two years. He was also NSF Fellow at Purdue University, US, in 1992. Dr Kothari, a recipient of the Most Active Researcher Award, has published 780 research papers in various national as well as international journals, conferences, has guided 45 PhD scholars and 65 M Tech students, and authored 42 books on power systems and other allied areas. He has delivered several keynote addresses and invitee lectures at both national and international conferences on electric energy systems. He has also widely popularised science and technology through 42 video lectures on YouTube with a maximum of 40,000 hits! Dr Kothari is a Fellow of the Indian National Academy of Engineering (FNAE); Fellow of Indian National Academy of Sciences [FNASc]; Fellow of Institution of Engineers (FIE) and Hon. Fellow, ISTE; and Fellow IEEE. His many awards include the National Khosla Award for Lifetime Achievements in Engineering (2005) from IIT Roorkee. The University Grants Commission (UGC), Government of India, has bestowed the UGC National Swami Pranavananda Saraswati Award (2005) on Education for his outstanding scholarly contributions. He is also a recipient of the Lifetime Achievement Award (2009) by the World Management Congress, New Delhi, for his contribution to the areas of educational planning and administration. His fields of specialization are Optimal Hydrothermal Scheduling, Unit Commitment, Maintenance Scheduling, Energy Conservation (loss minimization and voltage control), and Power Quality and Energy Systems Planning and Modelling.

Rakesh Singh Lodhi is Assistant Professor, Department of Electrical & Electronics Engineering, Oriental University, Indore (MP). He is a BE (Electrical & Electronics Engineering) from RGPV University and a postgraduate in Power Electronics from Shri Govindram Seksaria Institute of Technology and Science, Indore (MP). He is in academics and research since last seven years. He has guided many research projects on electrical motor drives at undergraduate as well as postgraduate levels. His research areas include Electrical Vehicles, Brushless DC Motor Drives, PWM Inverter, Control of Traction Drives, Industrial Drives and Performance of Power Electronics in Traction Drives. He has published more then 10 research papers in various national as well as international journals and conferences.

1

Introduction

1.1 INTRODUCTION

Automation control, motion control, and machine automation systems are used to improve manufacturing performance and flexibility. Engineering assistance with machine safety, energy efficiency, and motion control by braking system automation concepts, global application expertise, and support, reduce energy consumption. Improved workers' safety makes more effective use of new, integrated approaches to complex challenges of engineering. A production machine requires high precision movement, or just simple positioning. Integrated motor-drive technology implements this architecture, distributing functions such as motion control, safety and predictive maintenance to individual drives.

Nowadays modern controllers such as proportion controller, Norma L-2 controller deliver many more tasks and carry them out with great precision to incorporate both power electronic devices, microprocessors and microcontrollers. For many years, the motor controller was a box which provided the motor speed control and enabled the motor to adapt to variations in the load.

The tasks for modern control are:
- to control the transients and dynamics of the electrical or mechanical machine and apply its response to loads;
- to provide electronic commutation;
- to enable self-starting of the motor;
- to protect the motor and the controller from damage;
- to match the power from an available source to suit the requirements of the motor (voltage, frequency, number of phases); and
- to provide pure DC or AC power free from harmonics or interference so that it can be an integral part of a generator control system. Power conditioning could also be provided by a separate free standing module operating on any power source.

1.2 DRIVES

Motion control is required in a large number of industrial and domestic applications such as transportation systems, rolling mills, textile mills, machine tools, fans, motor pumps, robots, washing machines, etc. Systems employed for rotation control or speed control are called *drives*, and may employ any of the prime movers such as diesel or petrol engines, gas or steam turbines, steam engines, hydraulic motors, hydraulic turbines and electric motors, for supplying mechanical energy or electrical energy for motion control.

*Definition: The systems that are employed for motion control are called **drives**.*

1.3   CLASSIFICATION OF DRIVES

1.3.1   Based on the Mode of Operation

(1) Continuous duty drives: The motor works at a constant load for enough time to reach temperature equilibrium.

(2) Short time duty drives: The motor works at a constant load, but not long enough to reach temperature equilibrium. The rest periods are long enough for the motor to reach ambient temperature.

(3) Intermittent duty drives: Sequential, identical run and rest cycles with a constant load. Temperature equilibrium is never reached. Starting current has little effect on temperature rise.

1.3.2   Based on Means of Control

(1) Manual control drives: The drives with manual control can be either simple as a room fan or complicated as a push button starter.

(2) Semi-automatic control drives: Drives consisting of a manual device for giving a command and automatic device in response to a command.

(3) Automatic control drives: These drives have a control gear, without the manual device, are known as automatic control drives.

1.3.3   Based on Number of Machines

Individual electric drive: In this drive, each individual machine is driven by a separate motor. This motor also imparts motion to various parts of the machine.

- **Group drive:** This drive consists of a single motor, which drives one or more line shafts supported on bearings. The line shaft may be fitted with either pulleys and belts or gears, by means of which a group of machines or mechanisms may be operated. It is also sometimes called shaft drive.

- **Multi-motor electric drive:** In this drive, each operation of the mechanism is taken care of by a separate motor drive. In this drive system, there are several drives, each of which serves to actuate one of the working parts of the drive mechanisms.

1.3.4   Classification Based on Dynamics and Transients

(1) Independent transient period

(2) Dependent transient period

1.3.5   Classification Based on Methods of Speed Control

(1) Reversible and non-reversible independent constant speed.

(2) Reversible and non-reversible step speed control.

(3) Variable position control.

(4) Reversible and non-reversible smooth speed control.

1.4   ADJUSTABLE SPEED DRIVE (ASD) OR VARIABLE SPEED DRIVE (VSD)

It describes the equipment that is used to control the speed of machines. Many industrial processes such as assembly lines operate at different speeds for different products. Where process conditions demand adjustment of flow from a motor pump or fan, varying the speed of the drive may save energy compared with other techniques for flow control.

An adjustable speed drive is defined as the speed that may be selected from several different pre-set ranges. If the speed output can be changed without steps over a range, the drive is known as variable-speed drive.

1.4.1   Benefits of ASD

1. Smooth operation
2. Acceleration control
3. Different operating speeds
4. Compensate for changing process variables
5. Allow slow operation for setup purposes
6. Adjust the rate of production
7. Allow accurate positioning
8. Control torque or tension

1.5   FACTORS AFFECTING SELECTION OF DRIVES

(i) Steady-state operation needed: Behaviour of speed torque curve, speed regulation, speed range, efficiency, duty cycle, quadrant of operation, speed fluctuation, ratings.

(ii) Requirements of transient operation: Acceleration and deceleration, the performance of starting, motoring, breaking and reversing.

(iii) Source needed: Type of source, and its capacity, voltage magnitude, voltage fluctuations, power factor, harmonics and its effect on loads, ability to accept regenerated power.

(iv) Capital and running cost, maintenance needs, life

(v) Space and weight restriction

(vi) Reliability

1.6   CHOICE OF DRIVES

(i) Space requirement

(ii) For operation and maintenance, skilled personnel are needed

(iii) Obtainability of spare parts

(iv) Noise and other ecological considerations

(v) Consistency of drive like frequency of breakdown, up and down times

(vi) Operating cost, maintenance cost, etc.

(vii) Overall initial investment required

(viii) Load requirements, i.e., steady-state torque speed, characteristics of load, variable speed range required, duty cycle, whether the reversible process is required, a rating of the drive, etc.

(ix) Starting and braking requirements. The time required for accelerating to full speed, the time required for braking, the time required for reversal of the direction of rotation, etc.

1.7   APPLICATIONS OF DRIVES

There are many applications of drives. The selected ones are:

(1) Complicated metal cutting machine tools
(2) Paper making industry or paper mills
(3) Rolling machines or rolling mills
(4) Belt drives in mechanical systems
(5) Chain drives in mechanical systems
(6) Electrical motor drives in electrical systems
(7) Power electronics equipment in electrical systems
(8) Cement kilns or cement mills
(9) Aerospace actuators
(10) Automotive applications or automation system
(11) Robotic actuators
(12) Flexible manufacturing systems

1.8   LOAD EQUALIZATION

In the method of load equalization intentionally the motor inertia is increased by adding a flywheel on the motor shaft if the motor is not to be reversed. For effectiveness of the flywheel, the motor should have a prominent drooping characteristic so that on load there is a considerable speed drop.

In applications such as an electric hammer, pressing job, steel rolling mills, etc., load fluctuates widely within short intervals of time. In such drives, to meet the required load

Here
Tl is load torque;
Tlh is torque when current is maximum at th time; and
tl is time when current will be minimum.

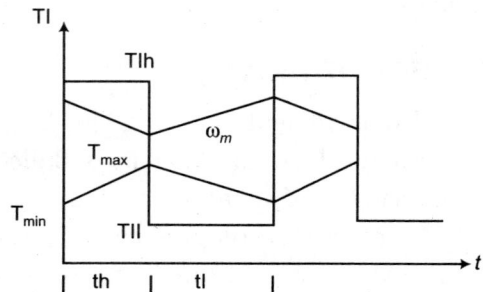

Fig. 1.1   Torque-time curve for load equalization.

the motor rating has to be high, or the motor would draw the pulse current from the supply. Such pulse current from the supply gives voltage fluctuations which affect the other load connected to it and affects the stability of the source. The above problem can be met by using a flywheel connected to the motor shaft for non-reversible drives. This is called load equalization. The moment of inertia and the mechanical time constant can be found out from the load equalization problem.

When an electric motor rotates, it is usually connected to a load which has a rotational or translational motion. The speed of the motor may be different from that of the load. To analyze the relation between the drives and loads, the concept of dynamics of electrical drives is introduced.

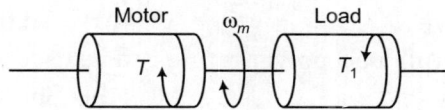

Fig. 1.2   Rotational system.

Motor load system

where

$\quad J$ = Polar moment of inertia of motor load

$\quad \omega_m$ = Instantaneous angular velocity

$\quad T$ = Instantaneous value of developed motor torque

$\quad T_1$ = Instantaneous value of load torque referred to motor shaft

Now, from the fundamental torque equation

$$T - T_1 = \frac{d}{dt}(J\omega_m) = J\frac{d\omega_m}{dt} + \omega_m\frac{dJ}{dt}$$

For drives with constant inertia

$$\frac{dJ}{dt} = 0$$

Therefore, $T = T_1 + J\dfrac{d\omega_m}{dt}$

So, the above equation states that the motor torque is balanced by load torque and a dynamic torque $J\,(d\omega_m/dt)$. This torque component is known as dynamic torque as it is only present during the transient operations. From this equation, we can determine whether the drive is accelerating or decelerating. Such as during accelerating motor supplies load torque and additional torque component essentially. So, the torque, balancing the dynamics of electrical braking is very helpful.

Multiple-Choice Questions

1. The systems employ for the motion control manually
   (a) Automatic control drives
   (b) Semi-automatic drives
   (c) Manual control drives
   (d) All of the above

2. Belt drives is an example of the system
   (a) Electrical system   (b) Mechanical system
   (c) Automation system   (d) Both (b) & (c)
3. There are not the type of drives
   (a) Computer   (b) Robotics
   (c) Car   (d) Electric train
4. The drive which operates on many machines can be controlled by a large number of motors
   (a) Group drive   (b) Individual drive
   (c) Multi-motor drive   (d) All the above
5. A motor of less than full load power rating can be used if the load is:
   (a) Continuous duty   (b) Short time duty
   (c) Intermittent periodic duty   (d) None of these
6. The consideration involved in the selection of the type of electric drive for a particular application depends on
   (a) Speed control range and its nature   (b) Starting torque
   (c) Environmental conditions   (d) All of the above

Answers

1. (c)   2. (d)   3. (a)   4. (a)   5. (b)
6. (d)

Exercise

1. Define drives and their classification.
2. What are variable speed drives?
3. Specify five examples of drive systems.
4. Write a short note on motion control system.
5. Comparison between the individual drive and multi-motor drive.
6. What are the advantages of adjustable speed drive?
7. Write an application of drives.
8. Describe the factors affecting the selection of drives.
9. Give characteristics of manual, semi-automatic and automatic drives.
10. What are the choices of drives?
11. Describe load equalization with a mathematical expression.

2

Electrical Drives

2.1 INTRODUCTION

Nowadays, modern power electronics and drives are used in electrical as well as mechanical industry. The power converter or power modulator circuits are used with electrical motor drives, providing either DC or AC outputs, and working from either a DC (battery) supply or from the conventional AC supply. Here we will highlight the most important aspects which are common to all types of drive converters. Although there are many different types of converters, all except very low-power ones are based on some form of electronic switching. The need to adopt a switching strategy is emphasized in the Wrist example, where the consequences are explored in some depth. We will see that switching is essential in order to achieve high-efficiency power conversion, but that the resulting waveforms are inevitably less than ideal from the point of view of the motor.

The thyristor DC drive remains an important speed-controlled industrial drive, especially where higher maintenance cost associated with the DC motor brushes (c.f. induction motor) is tolerable. The controlled (thyristor) rectifier provides a low-impedance adjustable DC voltage for the motor armature, thereby providing speed control.

2.1.1 Definitions of Electrical Drives

- An *electrical drive* can be defined as an electromechanical device for converting electrical energy into mechanical energy to impart motion to different machines and mechanisms for various kinds of process control.
- An *electrical drive* is an industrial system which performs the conversion of electrical energy into mechanical energy or vice versa for running and controling various processes.
- An *electrical drive* is defined as a form of machine equipment designed to convert electrical energy into mechanical energy and provide electrical control of the processes. The system employed for motion control is called an electrical drive.

2.2 ELECTRICAL DRIVES AND THEIR BLOCK DIAGRAM

An electrical drive system has the following components (**Fig. 2.1**).

1. Electrical machines and loads
2. Motor
3. Power modulator
4. Sources
5. Control unit
6. Sensing unit

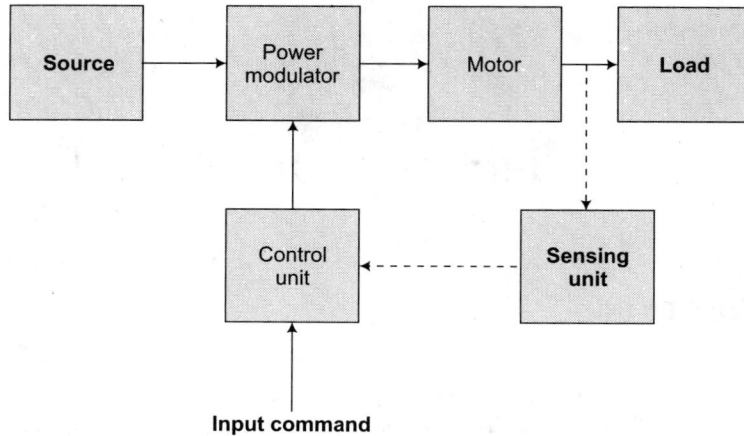

Fig. 2.1   Block diagram of electrical drives.

2.2.1   Electrical Machines and Loads

1. **DC Machines:** Shunt, series, compound, separately excited DC motors and switched reluctance machines.
2. **AC Machines:** Induction, wound rotor, synchronous, PM synchronous and synchronous reluctance machines.
3. **Special Machines:** Brushless DC motors, stepper motors, switched reluctance motors are used.

2.2.2   Load

The load is strictly W, kW or MW and as rightly points out VA, kVA, etc., but not resistance or inductance in itself. In other words, the load is neither the device connected nor its impedance but rather the power (or VA) drawn by that device.

2.2.3   Motor

It may be AC motor, brushless DC motor, permanent magnet DC motor, AC motor, stepper motor, switched reluctance motors, universal motor, synchronous motor, and induction motor.

2.2.4   Functions of Power Modulator or Power Converter

1. Modulates flow of power from the source to the motor in such a manner that motor is imparted speed-torque characteristics required by the load.
2. During transient operation, such as starting, braking and speed reversal.
3. It restricts source and motor currents within permissible limits.
4. It converts electrical energy of the source in the form of suitable to the motor.
5. It selects the mode of operation of the motor, i.e., motoring and braking.

2.2.5   Types of Power Modulators

In the electric drive system, the power modulators can be any one of the following:

1. Controlled rectifiers (AC to DC converters)
2. Inverters (DC to AC converters)
3. AC voltage controllers (AC to AC converters)
4. DC choppers (DC to DC converters)
5. Cycloconverters (frequency conversion)

2.2.6   Electrical Sources

Very low power drives are generally fed from single-phase sources. Rest of the drives are powered from a three-phase source. Low and medium power motors are fed from a 400 V supply. For higher ratings, motors may be rated at 3.3 kV, 6.6 kV, and 11 kV. Some drives are powered from the battery.

2.2.7   Control Unit

Control unit for a power modulator is provided in the control unit. It matches the motor and power converter to meet the load requirements.

2.2.8   Sensing Unit

1. **Speed sensing:** Speed can be sensed by using a tachometer. Wind speed can be sensed by anemometer similarly both speed and velocity can be measured by the speedometer.
2. **Torque sensing:** Magnetoelastic torque sensor is used in-vehicles applications on race cars, automobile, and aircraft.
3. **Position sensing:** Motion can be sensed through GPS, vibrometer, and rotary encoder.
4. **Current sensing and Voltage sensing** from lines or from motor terminals.
5. **Temperature sensing:** Thermistor is a device which is used for temperature measurement.

2.3   CLASSIFICATION OF ELECTRICAL DRIVES

There are two types of electrical drives **(Fig. 2.2).**

(1) DC Drive: It is further classified into two types:

  1. Non-regenerative DC drives: Non-regenerative DC drives are the most conventional type. In their most basic form, they are able to control motor speed and torque in one direction.
  2. Regenerative DC drives: Regenerative adjustable speed drives, also known as four-quadrant drives, are capable of controlling not only the speed and direction of motor rotation but also the direction of motor torque.

(2) AC Drive: It is further classified into two types:
    1. Induction motor drives
    2. Synchronous motor drives

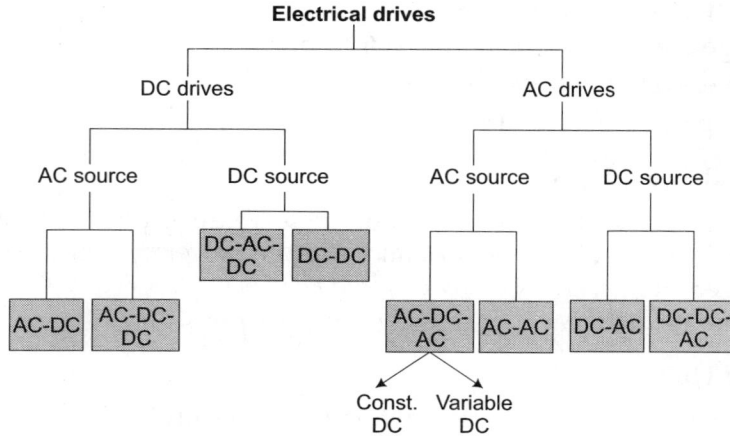

Fig. 2.2   Classification of electrical drives.

2.3.1   DC Drives

Electrical drives that use DC motors as the prime mover are called DC drives.

The DC drive is relatively simple and cheap, but DC motor itself is expensive. Due to the numerous disadvantages of DC motor it is losing popularity particularly in high power applications. For low power applications, the cost of DC motor plus drives is still economical. For servo application, DC drive is still popular because of good dynamic response and ease of control.

2.3.2   AC Drives

Electrical drives that use AC motors as the prime mover are called AC drives.

AC drives utilize a solid-state adjustable frequency inverter which adjusts frequency and voltage for varying the speed of an otherwise, conventional fixed speed AC motor. This is achieved through pulse-width modulation (PWM) of the drive output to the motors.

2.4   COMPONENTS OF ELECTRIC DRIVES

An electrical drive system uses electrical components. These components are made up of fewer mechanical wear parts, reducing the need to replace these parts, which result in lower operating costs for the electric drive system. Electric drive systems are very simple. They consist of an electric storage battery, a speed controller with the throttle, and a DC electric motor.

2.5   CHARACTERISTICS OF ELECTRIC DRIVES

An electric drive provides electrical retarding and reduces service brake wear. It also has many operational advantages. It includes the control of wheel slip and slide thus reducing the tire wears. The system delivers a smoother ride for the operator. The electric drive system enhancements include improved retarding grids, slip control algorithms, the latest in diagnostic and troubleshooting software and silencers.

2.6   MICROPROCESSOR BASED ELECTRICAL DRIVES

Microprocessor and microcontroller based electrical drives are DC motor drives, induction motor drives and traction motor drives.

2.6.1   Microprocessor Based DC Motor Drives

A microprocessor based control system can also be built where a phase controlled rectifier supplies a DC motor.

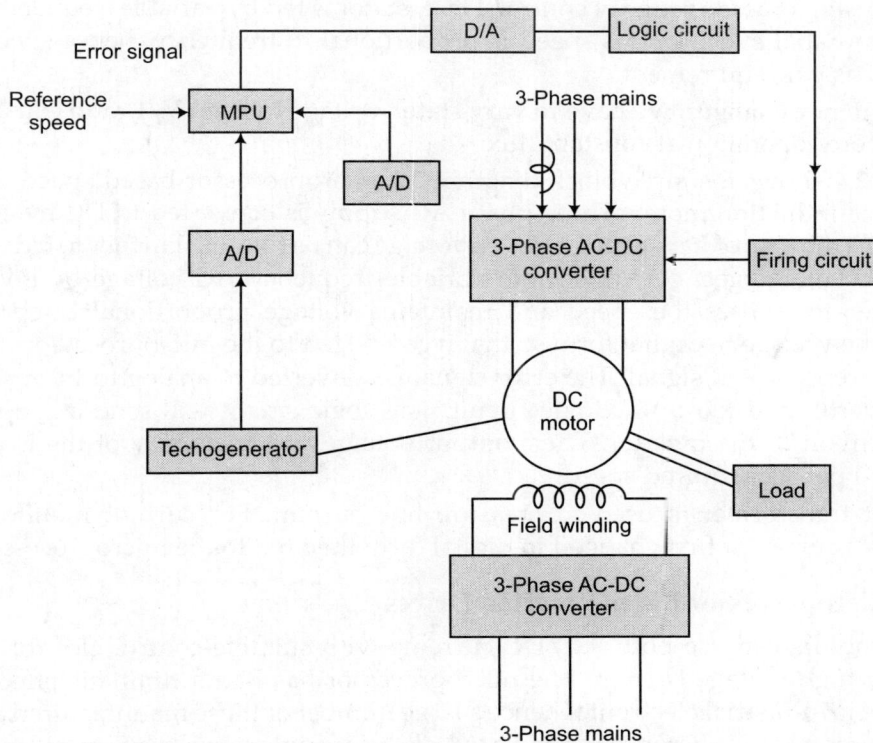

Fig. 2.3   Microprocessor based DC drives.

In **Fig. 2.3** the armature and field winding of DC motor are supplied from three-phase AC supply to AC-DC converter then it is fed to the armature and field winding. A techogenerator acts as a speed sensor, and it produces actual speed. Analog to Digital (A/D) converter changes the output voltage of speed sensor to digital form and feeds it to the microprocessor unit. A reference speed signal is also fed to the microprocessor unit. The error is converted to analog form by D/A converter that is connected to the logic circuit. It sends a signal to the firing circuit to adjust the firing angle of AC-DC converter feeding the armature. So the voltage applied to the armature is adjusted to control speed precisely. AC line current can be measured by the current transformer. A/D converter converts current signal into digital form connected to microprocessor unit for monitoring and controling the current.

2.6.2   Microprocessor Based Induction Motor Drives

Speed control method for three-phase induction motor:
• Stator voltage control
• Rotor control
• Stator voltage and frequency control

First two methods are used for limited range whereas stator voltage and frequency control is used for wide range of control. Here, stator is fed by variable frequency supply. As we know that synchronous speed is proportional to frequency, hence speed can be controlled as per requirement.

As frequency changes, we have to vary stator voltage to keep V/f ratio constant. This ensures motor operation at constant flux.

Figure 2.4 shows a simple block diagram for microprocessor based speed control of three-phase induction motor. Three-phase AC supply is converted to DC by controlled rectifiers. In the case of harmonics, this DC voltage can be filtered. Further to fed induction motor, DC voltage gets converted into variable frequency AC voltage by inverters. A techo-generator senses the speed and maintains voltage proportional to speed. A/D converts the voltage to digital form so that it can be fed to the microprocessor unit along with a reference speed signal. The error signal is converted to analog form by the use of D/A converter and fed to the logic circuit. This logic circuit will send a proper signal to firing circuit of the inverter. Thus, output voltage and frequency of the inverter are adjusted as per the required speed.

Current transformer is used to measure line current. For current monitoring and control, this current if first changed in digital form then fed to the microprocessor unit.

2.6.3   Microprocessor Based Traction Drives

Traction motors can use both AC/DC currents with suitable control electronics. They operate in high voltage current. The microprocessor has had a dramatic effect on PID controllers in industrial electronics hence a large number of those manufactured are based on microprocessors. Microprocessor based electric drive consists of an electric motor (motors), a transfer mechanism, an electrical energy converter, and a control system. The control system consists of a microcontroller with data connection interfaces, data channels (data networks), sensors and actuators. A generalized structure of the electric drive is shown in **Fig. 2.5.**

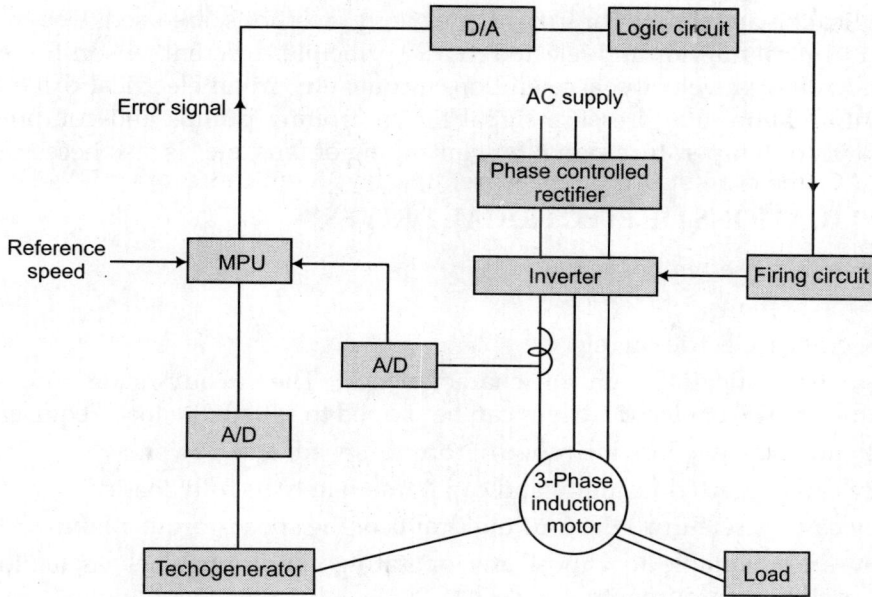

Fig. 2.4   Microprocessor based AC drives.

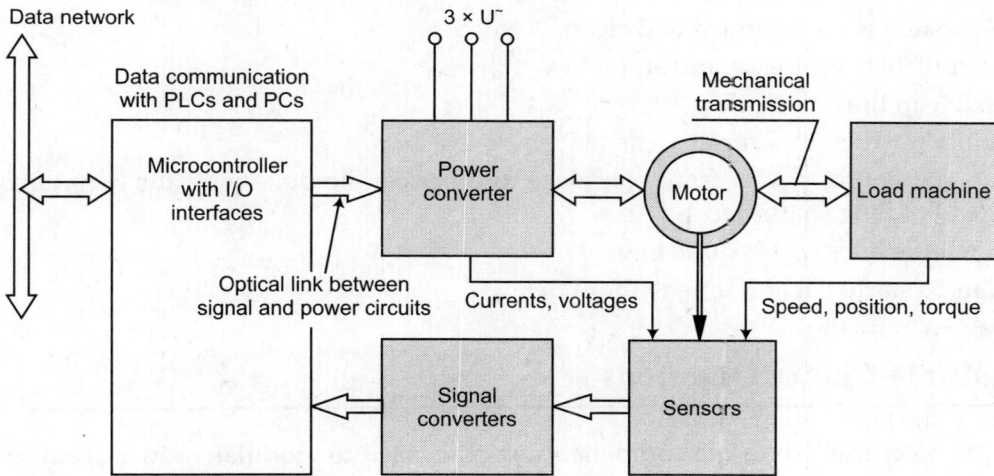

Fig. 2.5   Microprocessor based traction drives.

In general, the main task of the electrical drive is the motion control of mechanisms. An electrical drive is an automatic control system with a number of feedbacks where different automatic control principles, such as error driven feedback control, model-based control, logical binary control, or fuzzy logic control methods, are used. Depending on a particular technical solution and selected control principle, different sensors for measuring of currents, voltages, velocity, acceleration, torque, etc., in an electrical drive are used. Another information, like pressure signal for controlling pumps and compressors, air humidity and/or temperature signal for controlling of fans, etc., is also necessary.

2.7   APPLICATIONS OF ELECTRICAL DRIVES

Electric drives are used in boats,

1. traction systems,
2. lifts, cranes, electric car, etc.
3. They have flexible control characteristics. The steady state and dynamic characteristics of electric drives can be shaped to satisfy the load requirements.
4. They are available in wide range of torque, speed, and power.
5. They can be started instantly and can immediately be fully loaded.
6. They can operate in all the four quadrants of the speed-torque plane.
7. They are adaptable to almost any operating conditions such as explosive and radioactive environments.

2.7.1   Advantages of Electric Drives

- Cost is too low as compared to another system of the drive.
- The system is more simple and clean.
- The control is very easy and smooth.
- Flexible in the layout.
- Facility for remote control.
- Transmission of power from one place to other can be done with the help of cables instead of long shafts, etc.
- Its maintenance cost is quite low.
- It can be started at any time without delay.

Multiple-Choice Questions

1. In electrical drives the component which is used to modulate power from source to motor.
   - (a) Control unit
   - (b) Power modulator
   - (c) Sensing unit
   - (d) Control command
2. The power modulator which is used to convert fixed DC voltage to variable DC voltage.
   - (a) Controlled rectifier
   - (b) Un-controlled rectifier
   - (c) Chopper
   - (d) AC voltage controller

3. The electrical drives which are used above 70% in industry.
   (a) Induction motor drives   (b) Synchronous motor drives
   (c) DC drives   (d) (a) and (b)
4. The electrical drives which are less popular due to cost of motor in high power applications
   (a) DC drives   (b) AC drives
   (c) All the above   (d) None of the above
5. In power modulator frequency converter is known as:
   (a) Chopper   (b) Inverter
   (c) Cycloconverter   (d) Rectifier
6. Ward-Leonard controlled DC drives are generally used for
   (a) Light duty excavators   (b) Medium duty excavators
   (c) Heavy duty excavators   (d) All of the above

Answer

1. (b)   2. (c)   3. (a)   4. (a)   5. (c)
6. (c)

Exercise

1. Define electrical drives. What are its components?
2. Why do we require switching strategy in electrical drives?
3. What are the elements of electrical drive systems?
4. What are the different modes of operation of an electric drives?
5. Write a short note on electrical drives and give their classification.
6. State the merits and demerits of electrical drives. What are its applications?
7. Write a short note on:
   (a) Electrical source   (b) Load
   (c) Non-regenerative DC drives   (d) Power converter
8. Explain how motor speed can be controlled through electrical drives.
9. List of advantages offered by AC drives over DC drives.
10. Write and explain the components of a power modulator.
11. Compare non-regenerative DC drive and regenerative DC drive.
12. Why we are using power modulator in the electrical drive system?
13. List the advantages of electrical drives system.
14. Write and explain the working of power modulator with their classification.

15. Describe the working operation of microprocessor-based electrical drives.
16. Draw and explain the block diagram of microprocessor based DC drives.
17. Draw and explain the block diagram of microprocessor based AC drives.
18. Describe the working of A/D and D/A converter in microprocessor based drives.
19. Write a note on techo-generator.
20. Draw the basic structure of microcontroller based traction drives.

3

DC Drives

3.1 INTRODUCTION

During the nineteenth century, DC motors were used extensively to draw power direct from the DC source. The advent of thyristors capable of handling large current has revolutionized the field of electric power control. DC motor drives are used for many speed and position control systems where their excellent performance, ease of control and high efficiency are desirable characteristics. DC motor speed control can be achieved using switch mode DC-DC chopper circuits. For both mains-fed and battery supplied systems, power MOSFETs are the ideal switching devices for the converter stage. Power MOSFET's devices can easily be used in chopper circuits for low power and high power DC motor drives for a vehicle, industrial or domestic applications.

A high-performance motor drive system must have good dynamic speed command tracking and load regulating response. DC motors provide excellent control of speed for acceleration and deceleration. The power supply of a DC motor connects directly to the field of the motor which allows for precise voltage control, and is necessary for speed and torque control applications direct current (DC) motors have been widely used in many industrial applications such as electric vehicles, steel rolling mills, electric cranes, and robotic manipulators due to precise, wide, simple, and continuous control characteristics. Traditionally, rheostat armature control method was widely used for the speed control of low power DC motors. However, the controllability, low cost, higher efficiency, and higher current carrying capabilities of static power converters brought a major change in the performance of electrical drives.

3.2 DESCRIPTION OF DC MOTORS

A direct current (DC) motor is a fairly simple electric motor that uses electricity and a magnetic field to produce torque, which turns the motor. At its most simple, a DC motor requires two magnets of opposite polarity and an electric coil, this act as an electromagnet. The repellent and attractive electromagnetic forces of the magnets provide the torque that causes the DC motor to turn.

3.2.1 Main Parts of a DC Machine

The main parts of a 4-pole DC machine as shown in **Fig. 3.1.**

1. Stator: It is a stationary part of the machine. The field winding is placed on stator side. The main parts of stator are: (1) yoke or frame, (2) pole core, (3) pole shoe (4) lifting eye, and (5) feet.

2. **Rotor:** It is a rotationary or movable part of the machine. The armature winding is placed on the rotor side. The main parts of the rotor are: (1) armature core, (2) commutator, (3) shaft, and (4) armature winding.

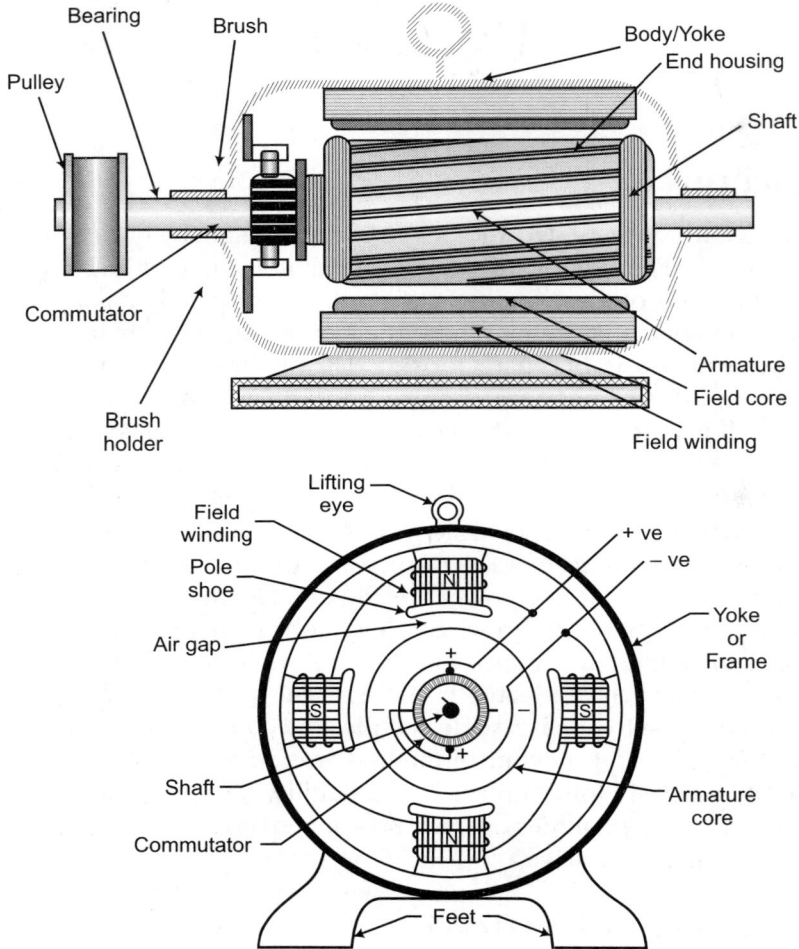

Fig. 3.1   Construction of DC machine.

3.2.2   Working Principle

DC machine is based on simple electromagnetism. A current-carrying conductor generates a magnetic field; when this is placed in an external magnetic field, it will experience a force proportional to the current in the conductor, and to the strength of the external magnetic field. As you are well aware of from playing with magnets as a kid, opposite (north and south) polarities, attract, while like polarities (north and north, south and south) repel. The internal configuration of a DC machine is designed to harness the magnetic interaction between a current-carrying conductor and an external magnetic field to generate rotational motion.

3.2.3   Classification of DC Motors

(1) Separately excited DC motor: The motor armature and field windings are supplied by different voltage sources. Both the armature and field winding currents can be adjusted conveniently. The motor speed can be easily controlled.

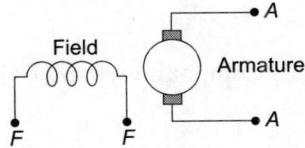

Fig. 3.2   Separately excited DC motor.

(2) Self-excited DC motor: The armature is supplied by an adjustable DC voltage source. The constant magnetic flux is generated by the permanent magnet.

Motor speed can be controlled by adjusting the armature current.

(1) DC series motor: Armature and field connected in a series circuit. Apply for high torque loads that do not require precise speed regulation. Useful for high breakaway torque loads.

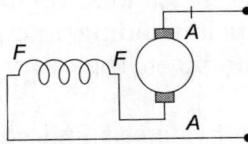

Fig. 3.3   DC series motor.

(2) DC shunt motor: Field coil in parallel (shunt) with the armature. Current through field coil is independent of the armature. The result will be excellent speed control. Normally operate in constant speed condition.

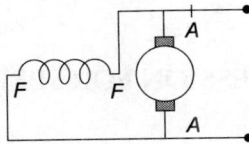

Fig. 3.4   DC shunt motor.

(3) Compound wound motor: Depending on whether the magnetic flux direction of the series field is aligned with that of the shunt field or not, four types of compound DC motors are there:

- Cumulative long compound.
- Cumulative short compound.
- Differential long compound.
- Differential short compound.

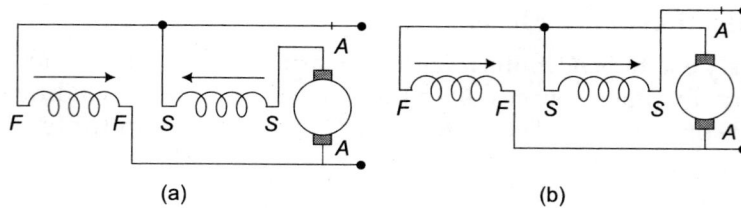

Fig. 3.5   Compound wound DC motor.

3.2.4   Losses in DC Motor

1. **Friction and Windage Loss:** These losses include bearing friction, brush friction, and windage. They are also known as mechanical losses. They are constant at a given speed but vary with changes in speed. Power losses due to friction increase as the square of the speed and those due to windage increase as the cube of the speed.

2. **Armature Copper Losses:** These are I^2R losses of the armature circuit, which includes the armature winding, commutator, and brushes. They vary directly with the resistance and as the square of the currents.

3. **Field Copper Losses:** These are I^2R losses of the field circuit which can include the shunt field winding, series field winding, interpole windings and any shunts used in connection with these windings. They vary directly with the resistance and as the square of the currents.

4. **Core Losses:** These are the hysteresis and eddy current losses in the armature. With the continual change of direction of flux in the armature iron, an expenditure of energy is required to carry iron through a complete hysteresis loop. This is the hysteresis loss. Also, since iron is a conductor and revolving in a magnetic field, a voltage will be generated. This, in turn, will result in small circulating currents known as eddy currents. They are reduced by using thin laminations, which are insulated from each other. Hysteresis and eddy current losses vary with flux density and speed.

3.3   MATHEMATICAL EXPRESSION FOR SPEED OF DC MOTORS

The equivalent circuit of DC motor:

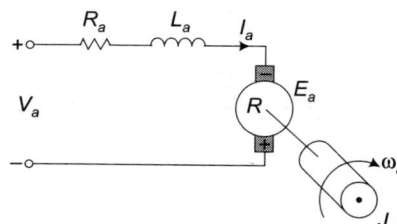

Fig. 3.6   DC motor equivalent circuit.

Figure 3.6 shows the mathematical term as:

V_a is armature voltage of DC motor T_e is electromagnetic torque of DC motor

i_a is armature current of DC motor ϕ is field flux of DC motor

R_a is armature resistance of DC motor K_t is constant

L_a is armature inductance of DC motor ω is speed of DC motor

E_a is back EMF of DC motor

Apply KVL in equivalent circuit of DC motor:

$$V_a = E_a + R_a\, i_a + L_a\, \frac{di_a}{dt} \tag{i}$$

From the torque equation of DC motor:

$$T_e = K_t\, \phi\, i_a \tag{ii}$$

From the EMF equation of DC motor:

$$E_a = K_t\, \phi\, \omega \tag{iii}$$

From equation (ii) find the value of i_a and put in equation (i). The term inductance is neglected in steady state.

So that
$$V_a = E_a + R_a \left(\frac{T_e}{K_t\, \phi} \right) \tag{iv}$$

From the equation of (ii) the value of E_a put in equation (iv)

$$V_a = K_t\, \phi\, \omega + R_a \left(\frac{T_e}{K_t\, \phi} \right) \tag{v}$$

Interchange the term V_a

So
$$K_t\, \phi\, \omega = V_a - R_a \left(\frac{T_e}{K_t\, \phi} \right) \tag{vi}$$

The speed is

$$\omega = \frac{V_a}{(K_t\, \phi)} - \left(\frac{T_e\, R_a}{(K_t\, \phi)^2} \right) \tag{vii}$$

According to equation (vii) speed is varied by armature voltage, field flux and armature resistance. Because:

$$\omega \propto V_a$$
$$\omega \propto R_a$$
$$\omega \propto \frac{1}{\phi}$$

Speed is directly proportional to armature voltage. So armature voltage control method is speed control of DC motor.

Speed is directly proportional to armature resistance. So speed control of DC motor with the variation of armature resistance.

Speed is inversely proportional to magnetic flux. So speed control of DC motor with the variation of flux or due to change of flux.

According to above term which shows there are three methods of speed control of DC motor.

• Armature voltage control method
• Armature resistance control method
• Magnetic field flux control method

3.3.1   Speed Control Methods of DC Motors

There are three methods of speed control of DC motor;

(1) Armature voltage control method
(2) Armature resistance control method
(3) Field flux control method

Two main types of control are available:

• Armature voltage control and
• Field flux control

These methods of control are combined to yield a wide range of speed control. Torque-speed characteristics are used to determine whether or not the performance is acceptable. Modern power converters constitute the power stage for variable-speed DC drives. These power converters are chosen for a particular application depending on a number of factors that include:

1. Cost
2. Available input power source
3. Harmonics
4. Power factor
5. Noise
6. Speed of response

An expression given in equation (vii) for mechanical speed in terms of armature terminal voltage and field flux was derived. From the above equation, it is evident that DC motor speed is directly proportional to the armature terminal voltage but inversely proportional to the field flux.

The armature voltage control method is applicable for all controlled rectifiers.

3.4   CONTROLLED RECTIFIER

The controlled rectifier can be operated as a single-phase or three-phase input. Output voltage contains low-frequency ripple which may require a large inductor inserted in armature circuit, in order to reduce the armature current ripple. A large armature current ripple is undesirable since it may be reflected in speed response if the inertia of the motor load is not large enough. The controlled rectifier has a low bandwidth. The average output voltage response to a control signal, which is the delay angle, is relatively slow. Therefore, controlled rectifier is not suitable for drives requiring a fast response, e.g., servo applications.

In terms of a quadrant of operations, a single-phase or a three-phase rectifier is only capable of operating in the first and the fourth quadrants – which is not suitable for drives requiring forward breaking mode. To be able to operate in all four quadrants, configurations using back-to-back rectifiers or contactors shown below must be employed.

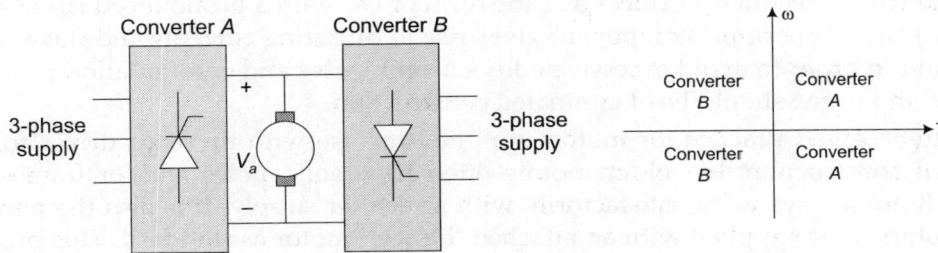

Fig. 3.7 Controlled rectifier and inverter combination.

3.5   THYRISTOR CONTROLLED DC DRIVES

Thyristor controlled drives operate at switching frequencies equal to supply frequency and are normally employed in high power applications. They are also used in low performance, low cost, and low power applications.

The speed of a DC motor can be controlled by controlling the DC voltage across its armature terminals. A phase-controlled thyristor converter can provide this DC voltage source. For a low-power drive, a single-phase bridge converter can be used, whereas for a high-power drive, a three-phase bridge circuit is preferred. The machine can be a permanent magnet or wound field type. The wound field type permits variation and reversal of field and is normally preferred in large power machines.

The main power circuit consists of a six-thyristor bridge circuit which rectifies the incoming AC supply to produce a DC supply to the motor armature. The assembly of thyristors, mounted on a heat sink, is usually referred to as the 'stack'. By altering the firing angle of the thyristors the mean value of the rectified voltage can be varied, thereby allowing the motor speed to be controlled.

Mathematical expression for controlled converters is:

1. Single-phase semi-controlled converter average output voltage:

$$\frac{V_m}{\pi}(V_a + \cos \alpha)$$

2. Single-phase fully controlled converter average output voltage:

$$V_a = \frac{2V_m}{\pi} \cos \alpha$$

3. Three-phase semi-controlled converter average output voltage:

$$V_a = \frac{3\sqrt{3}}{2\pi} V_m (1 + \cos \alpha)$$

4. Three-phase fully controlled converter average output voltage:

$$V_a = \frac{3\sqrt{3}\, V_m}{2\pi} \cos \alpha$$

The controlled rectifier produces a crude form of DC with a pronounced ripple in the output voltage. This ripple component gives rise to pulsating currents and fluxes in the motor, and in order to avoid excessive eddy-current losses and commutation problems, the poles and frame should be of laminated construction.

It is an accepted practice for motors supplied for use with thyristor drives to have laminated construction, but older motors often have solid poles and/or frames, and these will not always work satisfactorily with a rectifier supply. It is also the norm for drive motors to be supplied with an attached 'blower' motor as standard. This provides continuous ventilation and allows the motor to operate continuously at full torque even down to the lowest speeds without overheating.

Low power control circuits are used to monitor the principal variables of interest (usually motor current and speed), and to generate appropriate firing pulses so that the motor maintains constant speed despite variations in the load. The 'speed reference' (**Fig. 3.8**) is typically an analog voltage varying from 0 to 10 V, and obtained from a manual speed-setting potentiometer or from elsewhere in the plant.

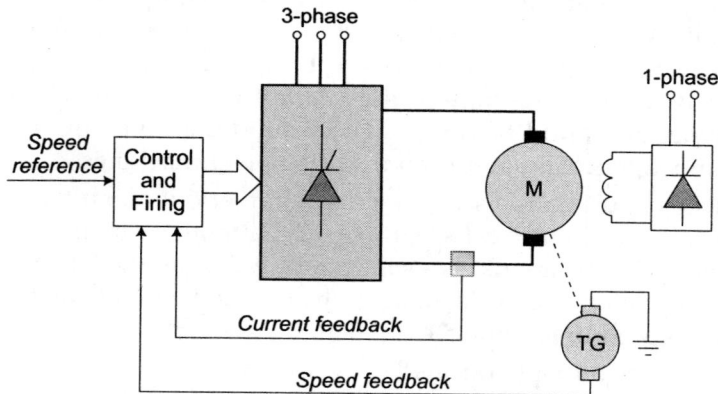

Fig. 3.8 Thyristor controlled DC drives.

The combination of power, control, and protective circuits constitutes the converter. Standard modular converters are available as off-the-shelf items in sizes from 0.5 kW up to several hundred kW while larger drives will be tailored to individual requirements. Individual converters may be mounted in enclosures with isolators, fuses, etc., or groups of converters may be mounted together to form a multi-motor drive.

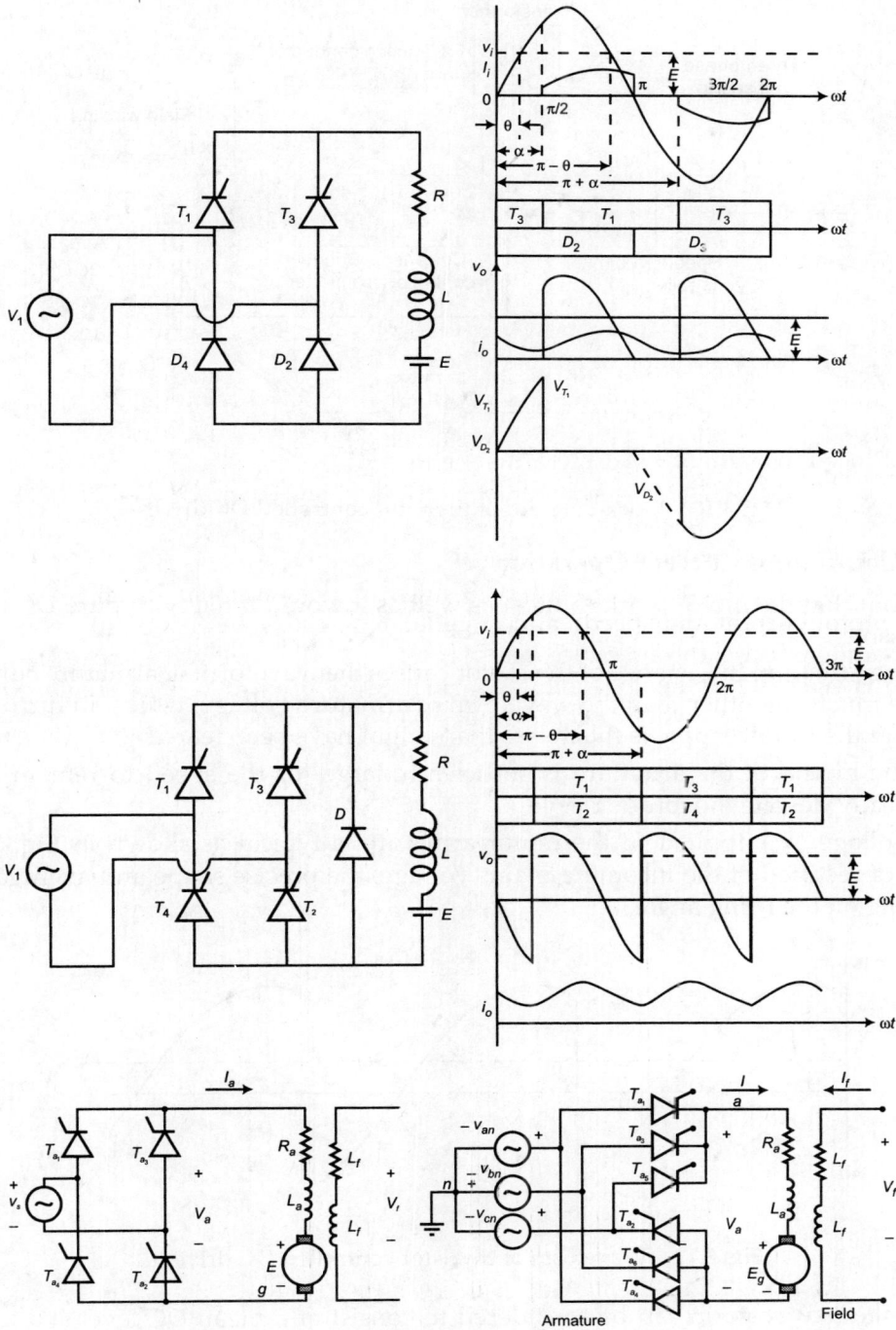

Fig. 3.9 Circuit diagrams and output voltage current waveforms of controlled converters fed DC drives.

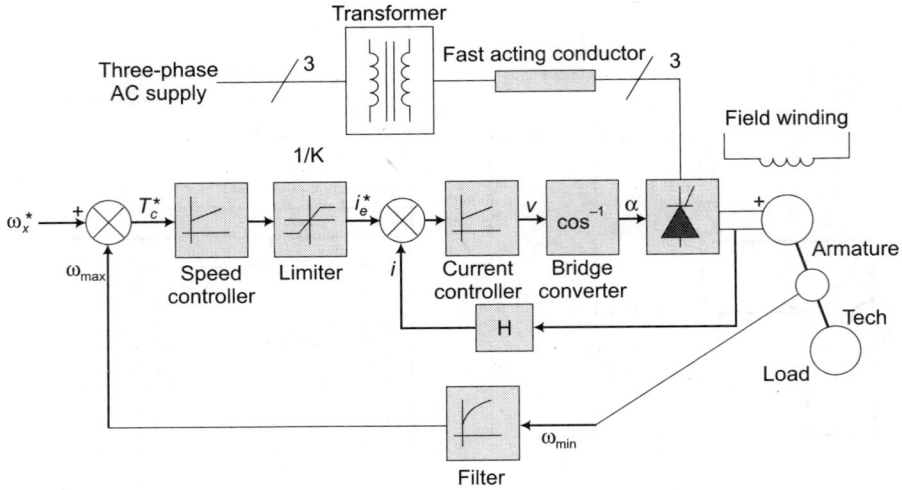

Fig. 3.10   Block diagram of thyristor controlled DC drives.

3.5.1   Continuous Current Operation

It turns out that the motor works almost as well as it would if fed with pure DC for two main reasons.

1. The armature inductance of the motor causes the waveform of armature current to be much smoother than the waveform of armature voltage, which, in turn, means that the torque ripple is much less than might have been feared.
2. The inertia of the armature is sufficiently large for the speed to remain almost steady despite the torque ripple.

The voltage (V_a) applied to the motor armature is typical as shown in **Fig. 3.11**, it consists of rectified of the incoming mains voltage, the precise shape and average value depending on the firing angle.

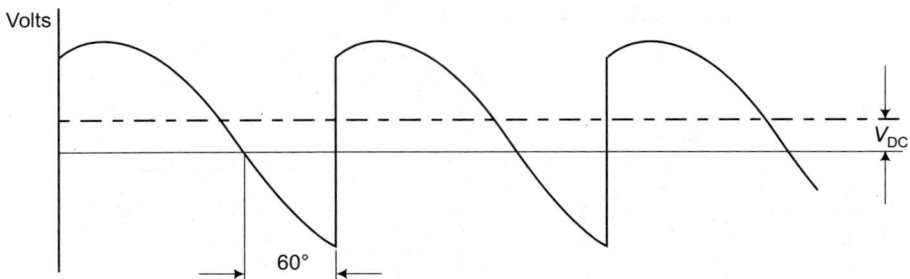

Fig. 3.11   Waveform of thyristor controlled DC drives.

The voltage waveform can be considered to consist of a mean DC level (V_{DC}), and a superimposed pulsating or ripple component which we can denote loosely as V_{AC}. The mean voltage V_{DC} can be altered by varying the firing angle, which also incidentally alters the ripple (i.e., V_{AC}).

The ripple voltage causes a ripple current to flow in the armature, but because of the armature inductance, the amplitude of the ripple current is small. In other words, the armature presents a high impedance to AC voltages. From which it can be seen that the current ripple is relatively small in comparison with the corresponding voltage ripple. The waveform of continuous current operation shown in **Fig. 3.12**.

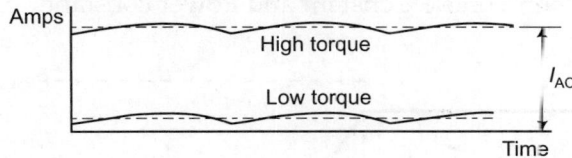

Fig. 3.12   Waveform of continuous current operation.

The average value of the ripple current is, of course, zero, so it has no effect on the average torque of the motor. There is nevertheless a variation in torque every half-cycle of the mains, but because it is of small amplitude and high frequency.

3.6   SPEED-TORQUE OPERATION

Armature voltage controlled DC drives are capable of providing rated current and torque at any speed between zero and the base (rated) speed of the motor. These drives use a fixed field supply and give motor characteristics as seen in **Fig. 3.13**. The motor output horsepower is directly proportional to speed (50% horsepower at 50% speed). The term constant torque describes a load type where the torque requirement is constant over the speed range.

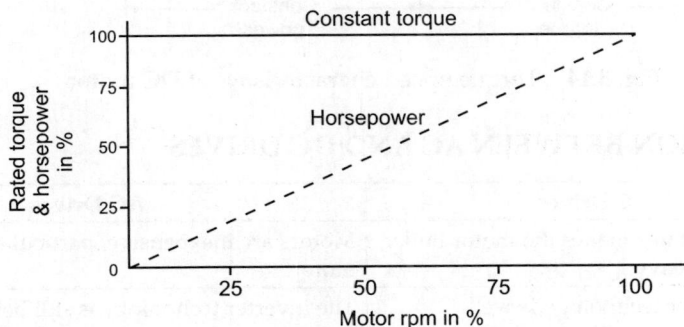

Fig. 3.13   Torque-speed characteristics.

$$\omega = \frac{V_a}{(K_t\,\phi)} - \left(\frac{T_e R_a}{(K_t\phi)^2}\right)$$

where K_t is a voltage constant. Increasing the load decreases the speed linearly. If the field current is varied within an appropriate range, constant speed can be maintained from no load to rated load. The power supply is in zero position its speed to the no load value. With constant field excitation, the motor in six steps until the motor current reaches rated value. The speed, torque, armature voltage and current, field voltage and current have been shown in **Fig. 3.14**. Reduce the load to zero. Start with the no-load speed at no load and gradually increase the load on the motor.

According to Fig. 3.14 with constant field excitation means I_f (field current) is constant. Six Step: Constant Torque, Constant Power, Constant Field Current, Above Base Speed torque will be decreased and field current also decreased. Armature current constant. So motor current reaches rated value. The speed torque, armature current, developed power and field current have shown in Fig. 3.14. In below waveform variations in T_d, P_d, I_f and I_a according to speed when Torque Constant and Power constant.

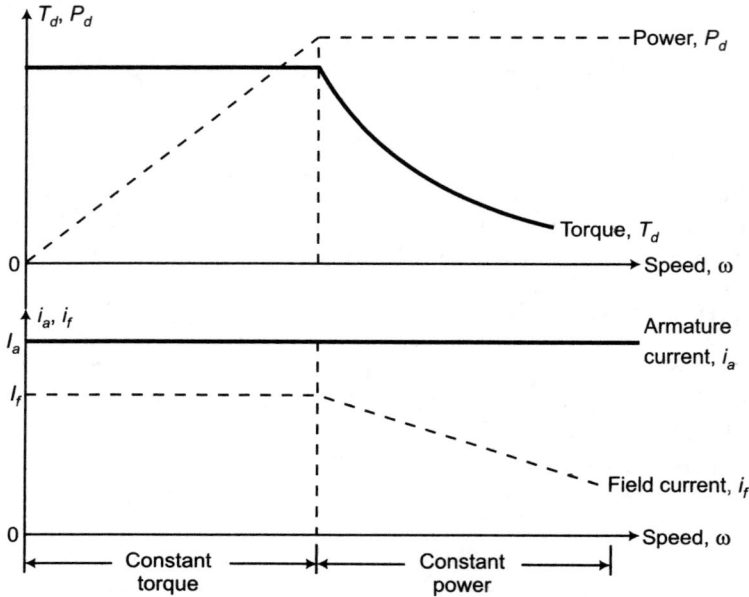

Fig. 3.14   Torque-speed characteristics of DC motor.

3.7   COMPARISON BETWEEN AC AND DC DRIVES

S. No.	DC Drives	AC Drives
1.	The commutator makes the motor bulky, costly and heavy.	Motors are inexpensive, particularly squirrel cage motor.
2.	The converter technology is well established. The power converter is simple and inexpensive.	The inverter technology is still being developed. The power circuit of the converter and its control are simple.
3.	Line commutation of the converter is used.	Forced commutation is used with induction motors. Sometimes machine commutation may be used with synchronous motors.
4.	Fast response and wide speed range smooth control.	With solid state converter, the speed range is wide. With conventional methods, it is stepped and limited.
5.	Small power/weight ratio.	Large power/weight ratio.
6.	Cost does not depend on the solid-state converter.	Solid-state converter employed also decides the cost.

Q.1   A 210 V, 850 rpm, 100 A separately excited DC motor has an armature resistance of 0.05 Ω. It is fed from a three-phase semi-controlled converter with an AC source of 230 V, 50 Hz. Assuming continuous conduction calculate:

(a) Firing angle for rated motor torque and 650 rpm

(b) Motor speed for α = 45 degrees and rated torque

Solution: Given values:

$$V_s = 210 \text{ V}, I_a = 100 \text{ A}, R_a = 0.05 \text{ Ω}$$
$$E_{b1} = V_s - I_a.R_a$$
$$E_{b1} = 210 - (100*0.05)$$
$$E_{b1} = 205$$

CASE 1:   At rpm = 650

$$\frac{E_{b1}}{E_{b2}} = \frac{N_1}{N_2}, N_1 = 850, N_2 = 650$$
$$E_{b2} = 156.7$$
$$V_a = \frac{3\sqrt{3}}{2\pi} V_m (1 + \cos \alpha)$$
$$E_{b2} = V_a - I_a.R_a,$$
$$156.7 = V_a - (100* 0.05)$$
$$V_a = 161.7 \text{ V},$$
$$V_a = \frac{3\sqrt{3}}{2\pi} V_m (1 + \cos \alpha),$$
$$V_m = 230\sqrt{2}$$
$$161.7 = \frac{3\sqrt{3}}{2\pi}.230\sqrt{2}.(1 + \cos \alpha)$$
$$\cos \alpha = -0.398$$

Firing angle at 650 rpm, α – 113.50°

CASE 2: For firing angle α = 45°

$$V_a = \frac{3\sqrt{3}}{2\pi} V_m (1 + \cos \alpha)$$
$$V_a = \frac{3\sqrt{3}}{2\pi}.230\sqrt{2}.(1 + \cos 45°)$$
$$V_a = 459.9 \text{ volt}$$
$$E_{b2} = V_a - I_a.R_a,$$
$$E_{b2} = 454.2,$$
$$\frac{E_{b2}}{E_{b1}} = \frac{N_2}{N_1},$$

Speed at Firing Angle α = 45° = 1883.2 rpm

Q.2   A 210 V, 850 rpm, 100 A separately excited DC motor has an armature resistance of 0.05 Ω. It is fed from a single-phase fully controlled rectifier with an AC source of 230 V, 50 Hz. Assuming continuous conduction calculate:

(a) Firing angle for rated motor torque and 650 rpm

(b) Firing angle for rated motor torque and – 500 rpm

(c) Motor speed for α = 45 degrees and rated torque

Solution: Given values:

$$V_s = 210 \text{ V}, I_a = 100 \text{ A}, R_a = 0.05 \text{ }\Omega$$
$$E_{b1} = V_s - I_a . R_a$$
$$E_{b1} = 210 - (100 * 0.05)$$
$$E_{b1} = 205$$

CASE 1: At rpm = 650

$$\frac{E_{b1}}{E_{b2}} = \frac{N_1}{N_2} , N_1 = 850, N_2 = 650$$

$$E_{b2} = 156.7$$

$$V_a = \frac{2V_m}{\pi} \cos \alpha,$$

$$E_{b2} = V_a - I_a . R_a,$$

$$156.7 = V_a - (100 * 0.05)$$

$$V_a = 161.7 \text{ V},$$

$$V_a = \frac{2V_m}{\pi} \cos \alpha,$$

$$V_m = 230\sqrt{2}$$

$$161.7 = \frac{230\sqrt{2} * 2 * \cos \alpha}{\pi}$$

$$\cos \alpha = 0.78$$

Firing angle at 650 rpm, α = 38.6°

CASE 2: At rpm = – 500

$$\frac{E_{b2}}{E_{b1}} = \frac{N_2}{N_1} , N_1 = 850, N_2 = -500$$

$$E_{b2} = -120.5$$

$$V_a = \frac{2V_m}{\pi} \cos \alpha,$$

$$E_{b2} = V_a - I_a . R_a,$$

$$120.5 = V_a - (100 * 0.05)$$

$$V_a = -125.5 \text{ V}$$

$$V_a = \frac{2V_m}{\pi}\cos\alpha,$$

$$V_m = 230\sqrt{2}$$

$$-125.5 = \frac{230\sqrt{2} * 2 * \cos\alpha}{\pi}$$

$$\cos\alpha = -0.60$$

Firing angle at – 500 rpm, $\alpha = 127.30°$

CASE 3: For firing angle $\alpha = 45°$

$$V_a = \frac{2V_m}{\pi}\cos\alpha,$$

$$V_a = \frac{2 * 230\sqrt{2}\cos45°}{\pi}$$

$$V_a = 146.4$$

$$E_{b2} = V_a - I_a.R_a,$$

$$E_{b2} = 141.4,$$

$$\frac{E_{b2}}{E_{b1}} = \frac{N_2}{N_1},$$

Speed at firing angle $\alpha = 45° = 586.2$ rpm

Q.3 A 200 V, 875 rpm, 150 A separately excited DC motor has an armature resistance of 0.06 Ω. It is fed from a single phase fully controlled rectifier with an AC source of 220 V, 50 Hz. Assuming continuous conduction calculates:

(a) Firing angle for rated motor torque and 750 rpm
(b) Motor speed for $\alpha = 160$ degrees and rated torque

Solution: Given values:

$$V_s = 200 \text{ V}, I_a = 150 \text{ A}, R_a = 0.06 \text{ }\Omega$$

$$E_{b1} = V_s - I_a.R_a$$

$$E_{b1} = 200 - (150 * .06)$$

$$E_{b1} = 191$$

CASE 1: At rpm = 750

$$\frac{E_{b1}}{E_{b2}} = \frac{N_1}{N_2}, N_1 = 875, N_2 = 750$$

$$E_{b2} = 163.7$$

$$V_a = \frac{2V_m}{\pi}\cos\alpha,$$

$$E_{b2} = V_a - I_a.R_a,$$

$$163.7 = V_a - (150 * 0.06)$$

$$V_a = 172.7 \text{ V},$$

$$V_a = \frac{2V_m}{\pi} \cos \alpha,$$

$$V_m = 220\sqrt{2}$$

$$172.7 = \frac{220\sqrt{2} * 2 * \cos \alpha}{\pi}$$

$$\cos \alpha = 0.87$$

$$\alpha = \mathbf{29.31°},$$

CASE 2: For firing angle $\alpha = 160°$

$$V_a = \frac{2V_m}{\pi} \cos \alpha$$

$$V_a = \frac{2 * 220\sqrt{2} \cos 160°}{\pi}$$

$$V_a = -186.12 \text{ V}$$

$$E_{b2} = V_a - I_a.R_a,$$

$$E_{b2} = -195.12,$$

$$\frac{E_{b2}}{E_{b1}} = \frac{N_2}{N_1},$$

$$\text{Speed} = -900.5 \text{ rpm}$$

Q.4   A 250 V, 850 rpm, 75 A separately excited DC motor has an armature resistance of 0.05 Ω. It is fed from a single-phase fully controlled rectifier with an AC source of 220 V, 50 Hz. Assuming continuous conduction calculate:

(a) Firing angle for rated motor torque and 675 rpm
(b) Firing angle for rated motor torque and -- 550 rpm
(c) Motor speed for $\alpha = 60°$ and rated torque

Solution: Given data:

(a)
$$V_a = 250 \text{ V}$$
$$I_a = 75 \text{ A}$$
$$N_1 = 850 \text{ rpm}$$
$$V_m = \sqrt{2} \text{ rms}$$
$$V_m = \sqrt{2} \times 250$$
$$V_m = 353.5 \text{ V}$$
$$E_b = V_a - I_a R_a$$
$$E_{b1} = 220 - 75 \times 0.05$$
$$E_{b1} = 216.52 \text{ V}$$
$$\frac{E_{b2}}{E_{b1}} = \frac{N_2}{N_1}$$

$$E_{b2} = \frac{N_2}{N_1} E_{b1}$$

$$E_{b2} = \frac{675}{850} \times 216.25$$

$$E_{b1} = 171.92 \text{ V}$$

$$V_a = E_b + I_a R_a$$

$$V_a = 171.92 + 75 \times 0.05$$

$$V_a = 88.75 \text{ V}$$

$$V_a = \frac{2 V_m \cos \alpha}{\pi}$$

$$88.75 = \frac{2 \times 353.5 \cos \alpha}{\pi}$$

$$\cos \alpha = \frac{88.75 \times 3.14}{2 \times 353.5}$$

$$\alpha = \cos^{-1} 0.394$$

$$\alpha = 66.7°$$

(b) for 550 rpm

$$\frac{E_{b2}}{E_{b1}} = \frac{N_2}{N_1}$$

$$E_{b2} = \frac{N_2}{N_1} E_{b1}$$

$$E_{b2} = \frac{550}{850} \times 216.25$$

$$E_{b2} = 140 \text{ V}$$

$$V_a = E_b + I_a R_a$$

$$V_a = 140 + 75 \times 0.05$$

$$V_a = 143.75 \text{ V}$$

$$V_a = \frac{2 V_m \cos \alpha}{\pi}$$

$$143.75 = \frac{2 \times 353.5 \cos \alpha}{\pi}$$

$$\cos \alpha = \frac{143.75 \times 3.14}{2 \times 353.5}$$

$$\alpha = \cos^{-1} 0.638$$

$$\alpha = 50.3°$$

(c) $\alpha = 60°$ motor speed $= ?$

$$V_a = \frac{2V_m \cos \alpha}{\pi}$$

$$V_a = \frac{2 \times \sqrt{2} \times 250 \cos 60}{\pi}$$

$V_a = 112.5 \text{ V}$

$E_b = V_a - I_a R_a$

$E_b = 112.5 - 75 \times 0.05$

$E_b = 136.25 \text{ V}$

$E_{rms} = 250 - 75 \times 0.05$

$E_{rms} = 246.25 \text{ V}$

$$N_2 = \frac{E_{b2}}{E_{b1}} \times N_1$$

$$N_2 = \frac{112.25}{246.25} \times 850$$

$N_2 = 388.59 \text{ rpm}$

Q.5 A 210 V, 900 rpm, 90 A separately excited DC motor has an armature resistance of 0.08 Ω. It is fed from a single-phase fully controlled rectifier with an AC source of 210 V, 50 Hz. Assuming continuous conduction calculates:

(a) Firing angle for rated motor torque and 600 rpm
(b) Firing angle for rated motor torque and – 600 rpm
(c) Motor speed for $\alpha = 55$ degrees and rated torque

Solution: Given data:

(a)

$V = 210 \text{ V}$

$I_a = 90 \text{ A}$

$N_1 = 900 \text{ rpm}$

$R_a = 0.08 \text{ Ω}$

$V_m = \sqrt{2} \text{ rms}$

$V_m = \sqrt{2} \times 210$

$V_m = 296.98 \text{ V}$

$E_b = V_a - I_a R_a$

$E_{b1} = 210 - 90 \times 0.08$

$E_{b1} = 202.8 \text{ V}$

$$\frac{E_{b2}}{E_{b1}} = \frac{N_2}{N_1}$$

$$E_{b2} = \frac{N_2}{N_1} E_{b1}$$

$$E_{b2} = \frac{600}{900} \times 202.8$$

$$E_{b2} = 135.2 \text{ V}$$

$$V_a = E_b + I_a R_a$$

$$V_a = 135.2 + 90 \times 0.08$$

$$V_a = 142.4 \text{ V}$$

$$V_a = \frac{2 V_m \cos \alpha}{\pi}$$

$$142.4 = \frac{2 \times 296.98 \cos \alpha}{\pi}$$

$$\cos \alpha = \frac{142.4 \times 3.14}{2 \times 296.98}$$

$$\alpha = \cos^{-1} 0.753$$

$$\alpha = 41.13°$$

(b) When speed is – 600 rpm

$$\frac{E_{b2}}{E_{b1}} = \frac{N_2}{N_1}$$

$$E_{b2} = \frac{N_2}{N_1} E_{b1}$$

$$E_{b2} = \frac{-600}{900} \times 202.8$$

$$E_{b2} = - 135.2 \text{ V}$$

$$V_a = E_b + I_a R_a$$

$$V_a = - 135.2 + 90 \times 0.08$$

$$V_a = - 128 \text{ V}$$

$$V_a = \frac{2 V_m \cos \alpha}{\pi}$$

$$- 128 = \frac{2 \times 296.98 \cos \alpha}{\pi}$$

$$\cos \alpha = \frac{- 128 \times 3.14}{2 \times 296.98}$$

$$\alpha = \cos^{-1} (- 0.67)$$

$$\alpha = 132.5°$$

(c) $\alpha = 55°$, motor speed = ?

$$V_a = \frac{2V_m \cos \alpha}{\pi}$$

$$V_a = \frac{2 \times \sqrt{2} \times 210 \cos 55}{\pi}$$

$$V_a = 108.44 \text{ V}$$

$$E_b = V_a - I_a R_a$$

$$E_b = 108.44 - 90 \times 0.08$$

$$E_b = 101.24 \text{ V}$$

$$N_2 = \frac{E_{b2}}{E_{b1}} \times N_1$$

$$N_2 = \frac{101.24}{202.8} \times 900$$

$$N_2 = 450 \text{ rpm}$$

Multiple-Choice Questions

1. When smooth and precise speed control over a wide range is desired, the motor preferred is
   (a) Synchronous motor
   (b) Squirrel cage induction motor
   (c) Wound rotor induction motor
   (d) DC motor

2. When quick speed reversal is a consideration, the motor preferred is
   (a) Synchronous motor
   (b) Squirrel cage induction motor
   (c) Wound rotor induction motor
   (d) DC motor

3. The ....motors, because of their inherent characteristics, are best suited for the rolling mills
   (a) DC
   (b) Slip ring induction
   (c) Squirrel cage induction
   (d) Single phase

4. In a constant power type load
   (a) Torque is proportional to speed
   (b) Torque is proportional to square of speed
   (c) Torque is inversely proportional to speed
   (d) Torque is independent of speed

5. To get speed higher than the base speed of the DC shunt motor
   (a) Armature resistance control is used
   (b) Field resistance control is used
   (c) Armature voltage control is used
   (d) None of these

6. The starting torque of a DC motor is independent of which of the following
   (a) Flux
   (b) Armature current
   (c) Flux and armature current
   (d) Speed

7. For a DC shunt motor which of the following is incorrect?
   (a) Unsuitable for heavy duty starting
   (b) Torque varies as armature current
   (c) Torque-armature current is a straight line
   (d) Torque is zero for zero armature current

8. In a separately excited DC motor, constant torque control of speed is achieved by varying
   (a) Armature voltage
   (b) Field current
   (c) Armature voltage
   (d) None of the above

9. In the armature control method, if the supply voltage V_a is decreased by a large amount, then the motor
   (a) speed gets reduced
   (b) gets overloaded
   (c) is constrained to work as a generator
   (d) quickly comes to a standstill

10. DC drives that use series motors are more advantageous than drives that employ separately excited motors because they facilitate
    (a) high starting torque
    (b) frequent starting
    (c) frequent torque overloads
    (d) all of the above

11. A three-phase full-wave controlled rectifier feeds a separately excited DC motor. The developed torque is 140 Nm, armature current is 60 A and speed (ω) is 65 radian per sec., the back emf of the motor will be nearly
    (a) 152 V
    (b) 102 V
    (c) 178 V
    (d) 194 V

12. In a DC drive supplied by a single-phase AC source and a rectifier, discontinuous conduction can be avoided by
    (a) adding an inductor in series with the armature of the DC motor
    (b) decreasing the frequency of the AC supply
    (c) adding a capacitor in series with the armature of the DC motor
    (d) increasing the frequency of the AC supply

13. A single-phase half-wave controlled rectifier has 400 sin 314 t as the input voltage and R as the load. For a firing angle of 60° for the SCR, the average output voltage is
    (a) $400/\Pi$
    (b) $300/\Pi$
    (c) $240/\Pi$
    (d) $200/\Pi$

14. A single-phase full-wave midpoint thyristor converter uses a 230/200 V transformer with a centre tap on the secondary side. The P.I.V. per thyristor is
    (a) 100 V
    (b) 141.4 V
    (c) 200 V
    (d) 282.8 V

15. In a single-phase full converter, for continuous conduction, each pair of SCRs conducts for
    (a) $\Pi - \alpha$ (b) Π
    (c) α (d) $\Pi + \alpha$

16. In a single-phase full converter, for discontinuous load current and extinction angle $\beta > \Pi$, each SCR conducts for
    (a) α (b) $\beta - \alpha$
    (c) β (d) $\alpha + \beta$

17. In a single-phase semi-converter, for continuous conduction, each SCR conducts for
    (a) α (b) Π
    (c) $\alpha + \Pi$ (d) $\Pi - \alpha$

Answers

1. (d)	2. (d)	3. (a)	4. (c)	5. (b)
6. (d)	7. (a)	8. (a)	9. (c)	10. (d)
11. (a)	12. (a)	13. (b)	14. (a)	15. (b)
16. (b)	17. (d)			

Exercise

1. What are DC drives?
2. Give some applications of DC drives.
3. What are the three types of speed control?
4. What is continuous and discontinuous conduction?
5. State the principle of phase control in AC-DC converter.
6. Write an expression for speed in DC motor.
7. Draw the speed torque characteristics of DC motor.
8. What are the uses of phase controlled rectifiers in DC drives?
9. Describe the working principle and operation of three-phase semi-controlled converter supplied to separately excited DC motor drive. Draw the o/p voltage and current waveforms; obtain the expression of angular velocity in terms of torque with torque-speed characteristics.
10. Discuss the advantages and disadvantages of converter control of separately excited D.C. motor.
11. Write a short note on thyristor controlled DC drives.
12. Draw and explain the speed-torque characteristics of a separately excited DC shunt motor giving its area of application.

13. A 240 V, 950 rpm, 14 A separately excited DC motor has armature circuit resistance and inductance of 2 ohms and 150 mH respectively. It is fed from a single-phase half controlled rectifier with an AC the source voltage of 250 V, 50 Hz Calculate:

   (i) Motor torque for firing angle $\alpha = 60°$ and speed = 600 rpm.

   (ii) Motor speed for $\alpha = 60°$ and $T = 20$ Nm

14. 1000 rpm DC shunt motor has an armature resistance of 0.05 ohm and a field resistance of 220 ohms the magnetization curve for the machine is given by table when current is 21 A and voltage 220 V.

   Field current (A) 0.2   0.4   0.6   0.8   1.0   1.2   1.4

   Emf at 1000 rpm

   50   100   150   190   219   235   245

15. A 220 V, 1500 rpm, 10A separately excited DC motor with armature resistance 2 ohms and armature load is 50 mH is fed from a single-phase fully controlled rectifier, when AC source voltage is 230 V 50 Hz. Calculate the no load speed, speed, and torque, when boundary between continuous and discontinuous conduction for $\alpha = 60$ and $\alpha = 45$.

16. A DC motor draws an armature current 50 amp. At 1000 rpm, and the speed DC shunt motor has a speed range 3:1 calculate the current at the speed of 3000 rpm when speed control is achieved by

   (1) Field flux control

   (2) Armature voltage control

   The motor whose load is

   (a) Load torque is constant

   (b) Zero load power is constant

4

Four-Quadrant Operations of DC Drives

4.1 INTRODUCTION

Most electrical motions operated by a battery use permanent magnet (PM) motors, since in this case no battery current needs to be spent for generating the magnetic field. Reversing the armature current reverses the direction of rotation.

The only exception is large vehicles, where high power motors are needed, beyond the capability of permanent magnets. In this case a 'separately excited' motor is used, mostly the shunt version which, as long as the field winding is fed at constant voltage, behaves like a PM motor (constant flux).

In practice, since the shunt motor may malfunction should the field current accidentally becomes zero, what is used is a variation of the shunt motor called 'compound'. A small series field winding is added to guarantee a minimum amount of field as long as there is armature current. An advantage of the shunt motor is that direction is reversed by reversing the field current, rather than the much larger armature current.

The product of speed and torque gives power developed by a motor. In 1st quadrant, developed power is positive. Hence, machine works like a motor supplying mechanical energy. Operation in 1st quadrant is, therefore, called forward motoring. In the 2nd quadrant, power is negative. Hence, machine works under braking, i.e., opposing the motion. 3rd and 4th quadrants can be identified as reverse motoring and braking respectively.

For constant torque operation, the change of stator winding is made from series-star to parallel-star while for constant horsepower operation the change is made from series-delta to parallel-star. Regenerative braking takes place during the changeover from higher to lower speeds. Positive motor torque is defined as the torque, which produces acceleration or the positive rate of change of speed in a forward direction. Positive load torque is negative if it produces deceleration. These comprise an electrical machine (AC or DC motor), power electronic converter or converters, and associated control circuitry. Modern drives are rarely connected directly to the power supply. A power electronic power-processing interface is normally employed.

4.2 LOAD TORQUE

Load torques are mainly classified into active and passive torque. Active load torque is the torque which has the potential to drive the motor under equilibrium condition, e.g., gravitational force, tension, compression, and torsion, etc. Passive load torque is the torque which always opposes the motion and changes its sign on the reversal of motion, e.g., friction, windage, cutting, etc. There are three components of load torque:

1. Friction torque
2. Windage torque
3. Mechanical torque.

4.2.1   State of Equilibrium

At constant speed, if the developed motor torque is equal to the sum of load torque and friction, then the drive is in its state of equilibrium.

4.2.2   Steady-State Stability

If the changes from one state of equilibrium to another take place too slowly to have the effects on the above factors, the stability conditions refer to steady-state stability.

4.2.3   Constant Torque Operation

The change of stator winding is made from series-star to parallel-star Constant horsepower operation: The change is made from series-delta to parallel-star. Regenerative braking takes place during the changeover from higher to lower speeds.

4.3   DC MOTOR CONTROLS

Direct current (DC) electric motors are often used in traction vehicle drive applications such as, electric locomotives or transit cars. In such applications, motive power is controlled by regulating motor current, typically by means of a control system employing a chopper. The chopper control system is essentially a controlled switching system connected in the energizing circuit of the motor armature so as to meter current to the motor by periodically opening and closing. The ratio of the closed time of the switching system to the sum of the closed time and the open time is the duty factor of the system. During the closed period of the chopper, the motor armature windings are connected to a power source through a path of relatively low resistance and current builds toward some peak value. During the open period of the chopper, the resistance in this energizing circuit is increased and armature current, circulating through a freewheeling diode, decays from the magnitude attained during the chopper closed time. In this manner, pulses of current are periodically applied to the motor and an average motor current is established. The average motor current tends to remain relatively constant due to the smoothing action of the circuit inductance. In general, the circuit inductance is sufficient to smooth the pulsating current and prevent jerking or lurching of the vehicle so long as the current pulses are supplied at relatively frequent rates, such as, 200 to 400 Hz.

An advantage of the chopper-controlled DC motor system is the relatively simple implementation of electrical braking. In electrical braking, the DC motor is operated as a generator with current generated by the armature windings being dissipated in a braking resistance (dynamic braking) or being forced back to the source (regenerative braking). The chopper control system operates in the braking mode in a manner similar to its operation in the driving mode, i.e., the braking torque is regulated by the chopper by controlling the average armature currents. In braking, however, the armature generated voltage may be allowed to be several times the magnitude of the source voltage in order to obtain the desired braking torque. Consequently, during the "OFF" time of the chopper, the peak voltage across the chopper may rise to a level several times as high as the source voltage. This braking characteristic necessitates the use of sophisticated and expensive components capable of withstanding such large applied voltages without destructive effect. One method which is commonly employed to avoid the necessity of using chopper components capable of withstanding such relatively high voltages is to connect the chopper in shunt with a first resistor which is serially interconnected with a plurality of additional resistors in the motor current path in a voltage dividing manner, thereby reducing the voltage impressed across the controlled switch in the chopper when turned off. In using this resistive voltage divider approach, it is apparent that some means must be provided for removing the additional series connected resistors from the current path when it is desired to increase the average current in order to maintain the desired level of braking torque.

4.4   CLOSED LOOP OPERATION OF DC MOTOR

In order to build a closed loop controller, you need some information about the rotation of the shaft like the number of revolutions executed per second, or even the precise angle of the shaft. This source of information about the shaft of the motor is called "feedback" because it sends back information from the controlled actuator to the controller. A closed loop controller will regulate the power delivered to the motor to reach the required velocity. If the motor is to turn faster than the required velocity, the controller will deliver

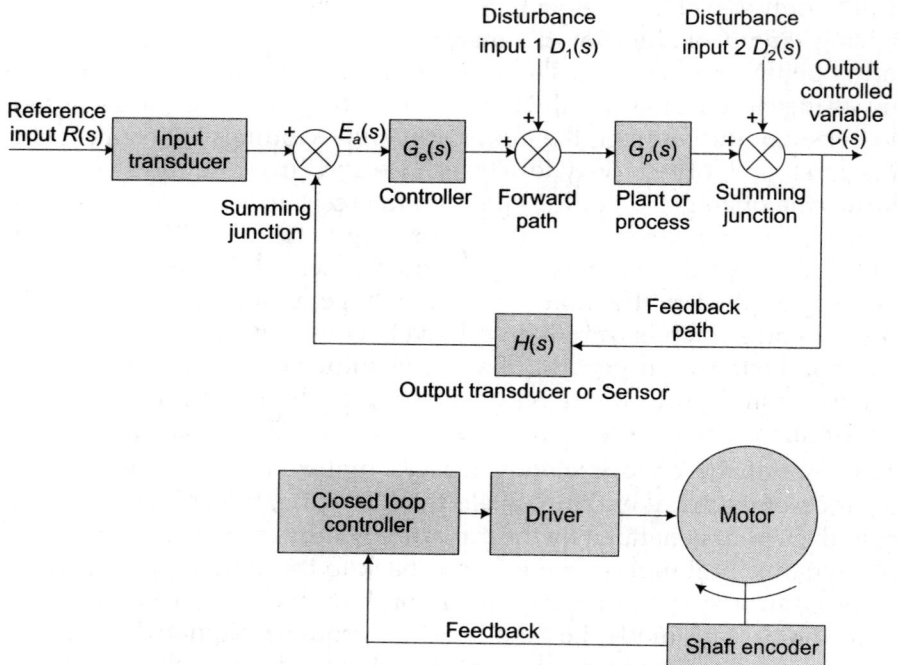

Fig. 4.1   Closed loop control of DC motor.

less power to the motor. Controlling the electrical power delivered to the motor is usually done by pulse width modulation.

A DC motor when driven by external means will generate a voltage. This voltage is also generated by the motor when it is supplied by a DC source. This voltage is proportional to the speed of the motor and is called the back emf. The best part is that the back emf is linearly proportional to the speed. To measure the back emf of the motor, we stop the PWM driver for a brief period. During this brief period, the motor coasts for some time when the current flows through the freewheeling diode. Once the energy stored in motor inductance is exhausted, the back emf builds up. This back emf is scaled to a suitable voltage using the potential divider. The back emf signal is then fed into the ADC input of the controller.

4.4.1   Advantages of Control Scheme

1. Cost effective. (This is the most important and the most sought after reason.)
2. Easy on hardware design.
3. Can be retrofitted on systems where motor does not come along with the encoder.
4. Small motors with gear used for hobby robotics usually run at high speed. (gear runs at low speed, gear reduction). Such motor can easily use this scheme.

4.4.2   Disadvantages of Control Scheme

1. Cannot be used to control the motor at extremely low speeds as the motor does not generate sufficient back emf.
2. Cannot be used in a system where torque ripple cannot be tolerated. Stopping the PWM for back emf measurement causes torque ripple.
3. If the motor is configured to run in both directions (using H-bridge), then the back emf measurement circuit becomes complicated (differential measurement).

4.5   TRANSFER FUNCTION MODEL OF DC MOTOR

From the equivalent circuit of DC motor

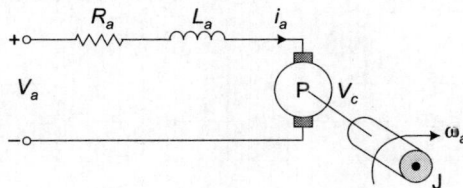

Fig. 4.2   Equivalent circuit of DC motor.

Steady-state equation of DC motor

$$V_a = R_a\, i_a + L_a\, \frac{di_a}{dt} + e_m \tag{1}$$

From the emf equation of DC motor: $e_m = K_e\, \dot{\theta}_m$ (2)

From the torque equation of DC motor: $T = K_t\, i_a$ (3)

For rotational motion

Electromechanical torque: $T = I_m\, \dot{\theta}_m + b\, \dot{\theta}_m$ (4)

where V_a = applied voltage i_a = armature current

       e_m = motor back emf K_e = motor voltage constant

       K_t = motor torque constant T = torque generated by motor

       I_m = equivalent moment of inertia reflected at the motor shaft

       b = equivalent viscous coefficient reflected at the motor shaft

Using Laplace transform, equations (1) to (4) can be written as

$$V_a(s) = R_a I_a(s) + L_a s I_a(s) + E_m(s)$$
$$E_m(s) = K_e s \Theta_m(s)$$
$$T(s) = K_t I_a(s)$$
$$T(s) = I_m s^2 \Theta_m(s) + bs \Theta_m(s)$$

Combining, we can write:

$$V_a(s) = \frac{R_a b}{K_t}\left[s(\tau_a s + 1)(\tau_m s + 1)\right]\Theta_m(s) + K_e s\, \Theta_m(s)$$

where $\tau_a = \dfrac{L_a}{R_a}$ and $\tau_m = \dfrac{I_m}{b}$ are the armature and the motor time constants respectively.

The transfer function for the armature-controlled DC motor is then:

$$\frac{\Theta_m(s)}{V_a(s)} = \frac{K_t/R_a b}{s\left[\tau_m \tau_a\, s^2 + (\tau_m + \tau_a)s + \left(\dfrac{K_e K_t}{R_a b} + 1\right)\right]}$$

The armature time constant τ_a is normally small compared to the motor time constant τ_m and the transfer function can be written as:

$$\frac{\Theta_m(s)}{V_a(s)} = \frac{K}{s(\tau s + 1)} \quad \text{or} \quad \frac{\omega_m(s)}{V_a(s)} = \frac{K}{(\tau s + 1)}$$

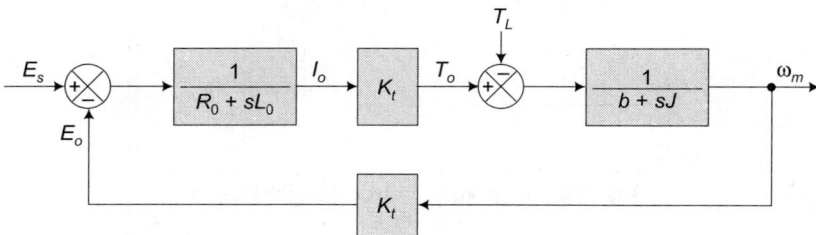

Fig. 4.3   Block diagram of DC motor.

It can be noted that for the armature-controlled DC motor, the transfer function between the output speed and the applied voltage is of the first order that is when a step voltage is applied to the motor, its speed will rise exponentially to a steady-state value. For angular position, however, the transfer function between the output position and the applied voltage is of 2nd order.

4.5.1 Closed-loop Position Control System

By taking the output shaft position and feeding it back to the input, a closed-loop position control system is obtained. This is indicated in **Fig. 4.4** with switch S_2 closed. In the figure, if switch S_1 is also closed, an inner velocity feedback loop is also present in addition to the outer position feedback loop. It can be shown that the closed-loop position control system using the armature-controlled DC motor, with or without the inner velocity feedback loop, is of 2nd order and is always stable.

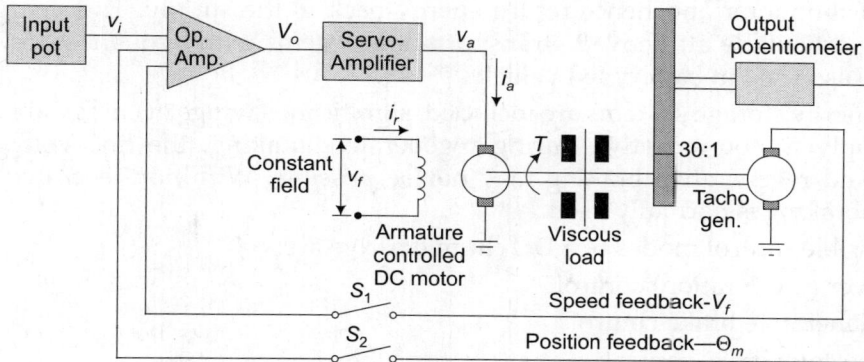

Fig. 4.4   Closed-loop position control system.

Switch S_1 and Switch S_2 both open	:	Open loop speed control system
Switch S_1 closed, and Switch S_2 open	:	Closed loop speed control system
Switch S_1 open, and Switch S_2 closed	:	Closed loop position control system without inner velocity feedback loop.
Switch S_1 and Switch S_2 both closed	:	Closed loop position control system with inner velocity feedback loop.

4.5.2 State-Space Modelling of DC Motor

The dynamic equations are arranged in state-space form as follows:

$$\begin{bmatrix} \dfrac{di_a}{dt} \\ \dfrac{d\omega_m}{dt} \end{bmatrix} = \begin{bmatrix} -\dfrac{R_a}{L_a} & -\dfrac{K_t}{L_a} \\ \dfrac{K_t}{J} & -\dfrac{B_L}{J} \end{bmatrix} \begin{bmatrix} i_a \\ \omega_m \end{bmatrix} + \begin{bmatrix} \dfrac{1}{L_a} & 0 \\ 0 & -\dfrac{1}{J} \end{bmatrix} \begin{bmatrix} V \\ T_L \end{bmatrix}$$

The roots of the system are evaluated as:

$$\lambda_1, \lambda_2 = \frac{-\left(\dfrac{R_a}{L_a} + \dfrac{B_L}{J}\right) \pm \sqrt{\left(\dfrac{R_a}{L_a} + \dfrac{B_L}{J}\right)^2 - 4\left(\dfrac{R_a B_L}{L_a J} + \dfrac{K_t^2}{L_a J}\right)}}{2}$$

The eigenvalues always have a negative real part indicating that the motor is always stable on the open-loop operation.

4.6   CHOPPER-CONTROLLED DC MOTOR DRIVES

Chopper DC drives are still widely used in traction applications. A DC-DC converter is connected between a fixed voltage DC source and a DC motor to vary the armature voltage. In addition to armature voltage control, the DC chopper can provide regenerative braking of the motor and hence return energy back to the supply. This energy saving feature is particularly attractive in transportation systems with frequent stops. Chopper drives are also used in battery fed vehicles.

When energy storage systems are included, significant savings in energy are achieved. If the supply is non-receptive during regenerative braking, the line voltage would increase, and regenerative braking may not be possible. With non-receptive supplies, rheostatic braking is normally used.

The possible control modes of a DC chopper drive are:
(a) Power (acceleration) control
(b) Regenerative brake control
(c) Rheostatic brake control
(d) Combined regenerative and rheostatic brake control

4.6.1   Four-quadrant Chopper Drive

Duty ratio may be varied either by keeping the switching frequency constant and adjusting the "ON" time or by keeping the "ON" time constant and adjusting the switching frequency. The first approach has the advantage of constant switching losses and a predetermined harmonic content. This translates into an optimal design for thermal management and filters. For a 4-quadrant DC-DC converter, output voltage control may be achieved either by using bipolar or unipolar voltage switching control. Bipolar voltage switching suffers from the following disadvantages:

• Doubling of switching losses as two switches are turned off instead of one.
• Doubling of rate of change of voltage across the load leading to higher electric field stresses and hence accelerated failure of motor windings,
• Higher rate of change of armature current (i.e., faster current dynamics) which may cause armature vibrations,
• Higher circulating currents in the armature circuit leading to increased copper losses.

4.6.2   Chopper

The thyristor in the circuit acts like a switch. The thyristor can be turned on or turned off as desired. When the thyristor is on, supply voltage appears across the load. When the thyristor is off, the voltage across the load will be zero. The output voltage and current waveforms are shown below.

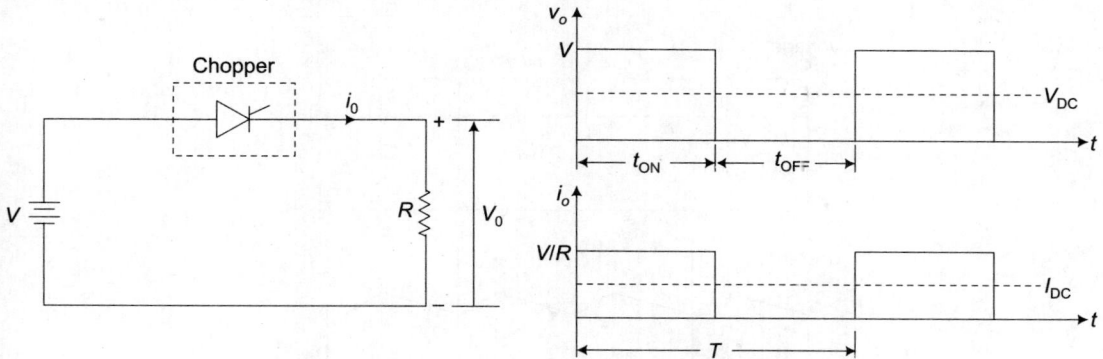

Fig. 4.5   Chopper circuits and their waveforms.

A single-switch chopper using a transistor, MOSFET or IGBT can only supply positive voltage and current to a DC motor, and is therefore restricted to the 1st quadrant motoring operation. When regenerative and/or rapid speed reversal is called for, more complex circuitry is required, involving two or more power switches, and consequently leading to increased cost.

Many different circuits are used and it is not possible to go into detail here, though it should be mentioned that the chopper circuit discussed in Chapter 2 only provides an output voltage in the range $0 < E$, where E is the battery voltage, so this type of chopper is only suitable if the motor voltage is less than the battery voltage. Where the motor voltage is greater than the battery voltage, a 'step-up' chopper using an additional inductance as an intermediate energy store is used.

4.6.3   Performance of Chopper-fed DC Motor Drives

We saw earlier that the DC motor performed almost as well when fed from a phase-controlled rectifier as it does when supplied with pure DC. The chopper-fed motor is, if anything, rather better than the phase-controlled because the armature current ripple can be less if a high chopping frequency is used.

Typical waveforms of armature voltage and current are shown in **Fig. 4.6(c)**. These are drawn with the assumption that the switch is ideal. A chopping frequency of around 100 Hz, as shown in **Fig. 4.6**, is typical of medium and large chopper drives, while small drives often use a much higher chopping frequency and thus have lower ripple current. As usual, we have assumed that the speed remains constant despite the slightly pulsating torque and that the armature current is continuous.

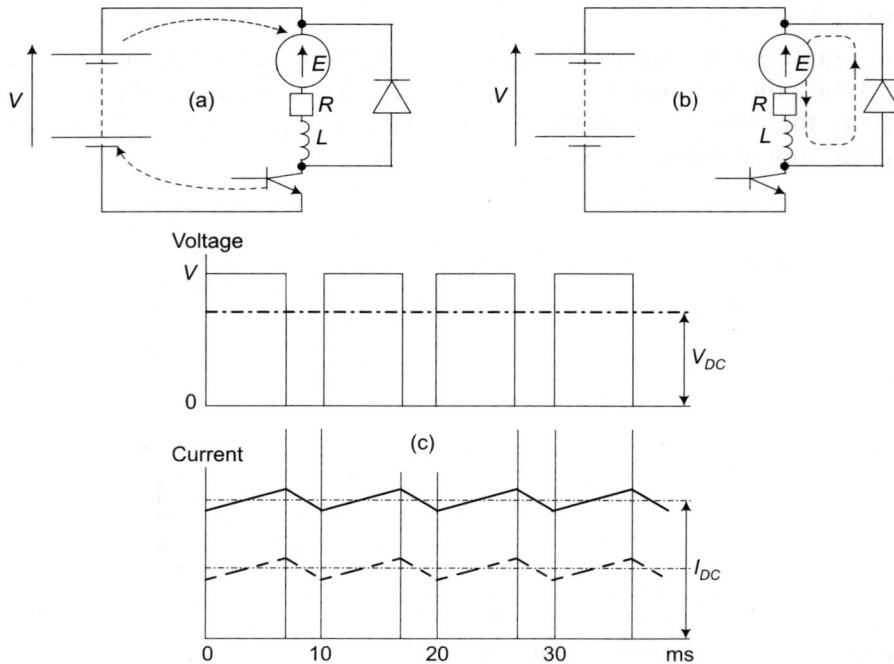

Fig. 4.6 Chopper fed DC drives with their waveforms.

The shape of the armature voltage waveform reminds us that when the transistor is switched on, the battery voltage V is applied directly to the armature, and during this period the path of the armature current is indicated by the dotted line **(Fig. 4.6(a))**. For the remainder of the cycle, the transistor is turned off and the current freewheels through the diode, as shown by the dotted line in **Fig. 4.6(b)**. When the current is freewheeling through the diode, the armature voltage is clamped at (almost) zero.

The speed of the motor is determined by the average armature voltage, (V_{DC}), which, in turn, depends on the proportion of the total cycle time (T) for which the transistor is on. Then the on and off times are defined as $T_{ON} = kT$ and $T_{OFF} = (1 - k)T$.

4.6.4 Chopper Drives Braking

The chopper controls the motor power in the driving mode. In the same way, it can be used to control regenerative power in the regenerative braking mode. The position of switch and diode will be interchanged. The function of the chopper in motoring and braking is the same. The on-off ratio of the chopper is regulated in a closed-loop control system to maintain the desired braking current. In regenerative braking, while the chopper is on, the motor terminals are shorted. The armature current builds up, and energy is stored in the reactor connected in series with the armature. When the chopper is off, armature current is forced into the supply. So, the energy stored in the reactor from the armature is thus released to the supply. If a series motor is used, an additional reactor in series with the armature may not be necessary. The series field winding stores energy during the chopper on interval and releases energy to the source while the chopper is off.

4.7 CONTROL STRATEGIES

PID control is a control strategy that has been successfully used for control of industrial processes over so many years. Simplicity, robustness, wide range of applicability and near optimal performance are some reasons behind the popularity of this control strategy. It has been observed that conventional tuning algorithms for this controller provide satisfactory performance as long as the system dynamics is exactly known. The output voltage can be controlled by varying the duty cycle.

Following are the methods for varying duty cycle:

1. Time Ratio Control (TRC)
2. Current-limit control

1. **Time ratio Control (TRC):** In this control scheme, time ratio T_{ON}/T (duty ratio) is varied. This is realized by two different ways called constant frequency system and variable frequency system as described below:

   Constant frequency system: In this scheme, on-time is varied but chopping frequency f is kept constant. Variation of T_{ON} means adjustment of pulse width, as such this scheme is also called pulse-width-modulation scheme.

   Variable frequency system: In this technique, the chopping frequency f is varied and either (i) on-time (T_{ON}) is kept constant or (ii) off-time (T_{OFF}) is kept constant. This method of controlling duty ratio is also called frequency-modulation scheme.

2. **Current-limit control (CLC):** In this control strategy, the ON and OFF of chopper circuit are decided by the previous set value of load current. The two set values are maximum load current and minimum load current. When the load current reaches the upper limit, the chopper is switched off. When the load current falls below the lower limit, the chopper is switched on. Switching frequency of chopper can be controlled by setting the maximum and minimum level of current. Current limit control involves feedback loop, the trigger circuit for the chopper is, therefore, more complex. PWM technique is the commonly chosen control strategy for the power control in chopper circuit.

4.7.1 Four-Quadrant Operation of DC Motor

Duty ratio may be varied either by keeping the switching frequency constant and adjusting the "ON" time or by keeping the "ON" time constant and adjusting the switching frequency. The first approach has the advantage of constant switching losses and a predetermined harmonic content. This translates into an optimal design for thermal management and filters. For a 4-quadrant DC-DC converter, output voltage control may be achieved either by using bipolar or unipolar voltage switching control.

Bipolar voltage switching suffers from the following disadvantages:

- Doubling of switching losses as two switches are turned off instead of one,
- Doubling of rate of change of voltage across the load leading to higher electric field stresses and hence accelerated failure of motor windings,
- Higher rate of change of armature current (i.e., faster current dynamics) which may cause armature vibrations,
- Higher circulating currents in the armature circuit leading to increased copper losses.

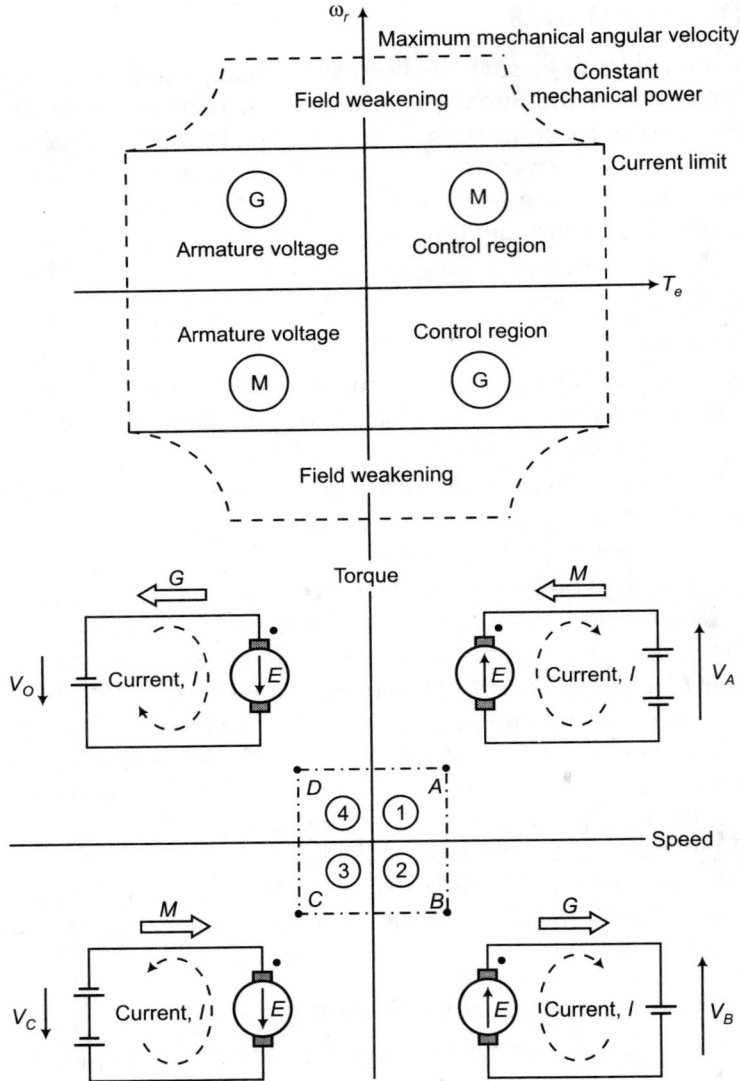

Fig. 4.7 Four-quadrant operation of DC motor.

4.8 DUAL CONVERTER

Dual converter is an electronic device or circuit made by the combination of two bridges. One of them works as a rectifier (converts AC to DC) and other bridge works as an inverter (converts DC into AC). Thus, an electronic circuit or device, in which two processes take place at the same time, is known as a dual converter. A dual converter may be single-phase or three-phase device. The difference between single-phase and three-phase dual converter is just that in three-phase we use three-phase rectifier at the first stage while in single-phase dual converter we make use of single-phase rectifier circuit at the first bridge.

4.8.1   Three-Phase Dual Converter

In three-phase dual converter, we make use of three-phase rectifier which converts three-phase AC supply to DC. The rest of the process is same and same elements are used. The output of three-phase rectifier is fed to a filter and after filtering pure DC is fed to load. At last the supply from the load is given to the last bridge, that is, the inverter. It converts DC into three-phase AC which appears at the output.

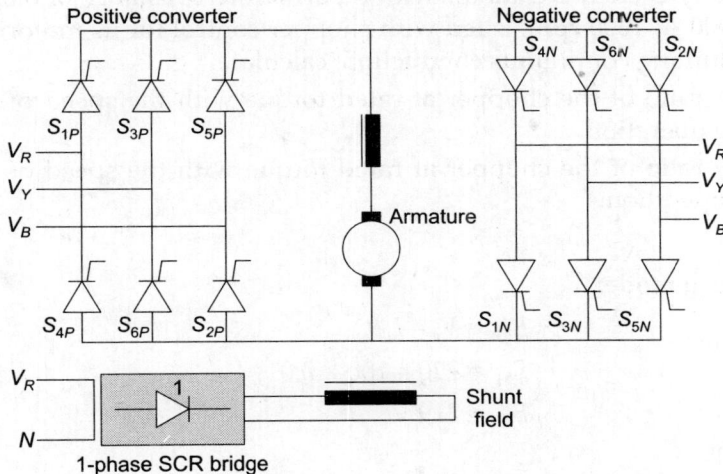

Fig. 4.8   Three-phase dual converter.

4.8.2   Applications of Dual Converter

Dual converters are mostly used in industries where we require reversible DC. Generally, dual converters are used for speed control of DC motors, etc.

Q.1   A 210 V, 25 A, 1500 rpm, DC motor having an armature resistance of 3 ohms is controlled by chopper. The chopping frequency is 500 Hz and input voltage is 230 V. Calculate the duty ratio of a vector torque 1.5 times the rated torque at 800 rpm.

Solution:

$$E_{b1} = V_a - I_a R_a$$
$$E_{b1} = 210 - 25 \times 3$$
$$E_{b1} = 135 \text{ V}$$

We know that,

$$\frac{E_{b1}}{E_{b2}} = \frac{N_1}{N_2}$$
$$\frac{135}{E_{b2}} = \frac{1500}{800}$$

$$E_{b2} = 72 \text{ V}$$
$$\delta V_a = E_{b2} + I_a R_a$$
$$\delta V_a = 72 + 25 \times 3$$
$$\delta E_a = 147$$

We know that,

$$\delta = V_a/E_a$$
$$= 210/147$$
$$\delta = 1.42$$

Q.2 A separately excited DC motor with an armature resistance of 0.01 ohm with DC supply 220 V, 100 A, 1000 rpm, is fed with chopper control for its motoring and braking operations. Assuming continuous conduction, calculate:

(i) The duty ratio of the chopper at rated torque with the speed of 500 rpm for its motoring operation.

(ii) The duty ratio of the chopper at rated torque with the speed of 500 rpm for its braking operation.

Solution:

(1) For motoring mode:

$$E_{b1} = V_a - I_a R_a$$
$$E_{b1} = 220 - 100 \times 0.01$$
$$E_{b1} = 219 \text{ V}$$

We know that,

$$\frac{E_{b1}}{E_{b2}} = \frac{N_1}{N_2}$$

$$\frac{219}{E_{b2}} = \frac{1000}{500}$$

$$E_{b2} = 109.5 \text{ V}$$
$$\delta V_a = E_{b2} + I_a R_a$$
$$\delta V_a = 109.5 + 100 \times 0.01$$
$$\delta E_a = 110.5$$

We know that,

$$\delta = V_a/E_a$$
$$= 220/110.5$$
$$\delta = 1.99$$

(2) For braking mode:

$$E'b_1 = V_a + I_a R_a$$
$$E'b_1 = 220 + 100 \times 0.01$$
$$E'b_1 = 221 \text{ V}$$

We know that,

$$\frac{E_{b1}}{E_{b2}} = \frac{N_1}{N_2}$$

$$\frac{221}{E'_{b2}} = \frac{1000}{500}$$

$$E'b_2 = 110.5 \text{ V}$$
$$\delta V_a = E_{b2} + I_a R_a$$
$$\delta V_a = 110.5 + 100 \times 0.01$$
$$\delta E_a = 111.5$$

We know that,

$$\delta = V_a/E_a$$
$$= 220/111.5$$
$$\delta = 1.97$$

Multiple-Choice Questions

1. A DC chopper is used for regenerative braking of a separately excited DC motor. The DC supply voltage is 400 V. The motor has $R_a = 0.2$ ohm $K_m = 1.2$ V-s/rad. The average armature current during regenerative braking is kept constant at 300 A with negligible ripple. For a duty cycle of 60% for a chopper, power returned to DC supply:
   (a) 90 kW   (b) 85 kW
   (c) 27 kW   (d) 100 kW

2. The ratio of turn on time to total time in chopper is known as:
   (a) Duty time   (b) Duty cycle
   (c) Chopping frequency   (d) Duty ratio

3. To save the energy during braking
   (a) Dynamic braking is used   (b) Plugging is used
   (c) Regenerative braking is used   (d) Mechanical braking is used

4. The advantage of electric braking is
   (a) It is instantaneous
   (b) More heat is generated during braking
   (c) It avoids wear of track
   (d) Motor continues to remain loaded during braking

5. The equilibrium speed of a motor load system is obtained
   (a) When motor torque equals the load torque
   (b) When motor torque is less than the load torque

(c) When motor torque is more than the load torque

(d) None of these

6. Which of the following is preferred for automatic drives?

(a) Synchronous motors

(b) Squirrel cage induction motor

(c) Ward Leonard controlled DC motors

(d) Any of the above

Answers

1. (b)   2. (d)   3. (c)   4 (c)   5 (a)

6. (c)

Exercise

1. Write a short note on control scheme of DC motor.
2. Define load torque?
3. Drive the expression for the transfer function of DC motor using 4-quadrant operations?
4. Drive the expression for a dynamic model of DC motor.
5. Write a short note on time ratio control and current limit control.
6. Draw the block diagram of a DC motor.
7. Define a chopper. Write and explain about the waveform of chopper circuit.
8. Write down the performance characteristics of chopper fed DC drives.
9. Explain the 4-quadrant operation of a DC motor.
10. Draw and explain closed loop control of a DC motor.
11. Describe the working of three-phase dual converter fed DC drives.
12. Draw and explain the multi-quadrant operation of chopper fed DC drives.
13. Describe time ratio control and current limit control of a chopper.
14. Describe the performance of a chopper fed DC drives in two-quadrant operation.

5

Induction Motor Drives

5.1 INTRODUCTION

For industrial and mining applications, three-phase AC induction motors are the prime movers for the vast majority of machines. These motors can be operated either directly from the mains or from adjustable frequency drives. In modern industrialized countries, more than half of the total electrical energy is converted into mechanical energy through AC induction motors. Applications for these motors cover almost every stage of manufacturing and processing. Applications also extend to commercial buildings and the domestic environment. They are used to drive pumps, fans, compressors, mixers, agitators, mills, conveyors, crushers, machine tools, cranes, etc. It is not surprising to find that this type of electric motor is so popular when one considers its simplicity, reliability, and low cost. In the last decade, it has become an increasingly common practice to use three-phase squirrel cage AC induction motors with variable voltage variable frequency (VVVF) converters for variable speed drive (VSD) applications. To clearly understand how the VSD system works, it is necessary to understand the principles of operation of this type of motor.

Although the basic design of induction motors has not changed much in the last 50 years. However, modern insulation materials, computer-based design optimization techniques, and automated manufacturing methods have resulted in motors of smaller physical size and lower cost per kW. International standardization of physical dimensions and frame sizes means that motors from most manufacturers are physically interchangeable, and they have similar performance characteristics. The reliability of squirrel cage AC induction motors, compared to DC motors, is higher.

The only parts of the squirrel cage motor that can wear are the bearings. Slip rings and brushes are not required for this type of construction. Improvements in modern pre-lubricated bearing design have extended the life of these motors.

Although single-phase AC induction motors are quite popular and common for low-power applications up to approx 2.2 kW, these are seldom used in industrial and mining applications. Single-phase motors are more often used for domestic applications.

To understand how an induction motor operates, we must first unravel the mysteries of the rotating magnetic weld. We shall see later that the rotor is effectively dragged along by the rotating weld, but that it can never run quite as fast as the weld. When we want to control the speed of the rotor, the best way is to control the speed of the weld. Our look at the mechanism of the rotating weld will focus on the stator windings because they act as the source of the flux. In this part of the discussion, we will ignore the presence of the rotor conductors. This makes it much easier to understand what governs the speed of rotation and the magnitude of the weld, which are the two factors that usually influence the motor behaviour.

Having established how the rotating weld is set up, and on what its speed and strength depend, we move on to examining the rotor, concentrating on how it behaves when exposed to the rotating weld and discovering how the induced rotor currents and torque vary with rotor speed. In this section, we assume again for the sake of simplicity – that the rotating flux set up by the stator is not influenced by the rotor. Finally, we turn our attention to the interaction between the rotor and stator, verifying that our earlier assumptions are well justified. Having done this, we are in a position to examine the 'external characteristics' of the motor, i.e., the variation of motor torque and stator current with speed. These are the most important characteristics from the point of view of the user.

5.2   HISTORY OF INDUCTION MOTOR DRIVE

The induction motor (as well as the synchronous and split-phase motors) was developed by Nikola Tesla in 1924 and has the endearing characteristic that it can be run by direct connection to a three-phase power source. The motor speed is directly proportional to the applied frequency and is determined by the formula $n = 120\ f/p$ where n is the synchronous speed of the motor in rpm, f is the frequency of power applied and p is the number of poles on the rotor. Therefore, a 2-pole induction motor running at 60 Hz will run at 3600 rpm synchronous speed less the slip required to produce the induction effect at full load. This slip is variable depending on the motor design but for the "standard" NEMA design B motor it is 3 to 5% making the typical 2-pole motor run at 3500 rpm at full load at 60 Hz. Soon after the AC motor was developed, the idea of varying the speed was considered and the only practical way of doing this at the time was to provide the motor with a variable frequency obtained by using a DC motor turning an AC alternator which allowed a variable frequency. This was done on a wide range of applications in the 50s, 60s, and 70s. Since the much simpler Ward-Leonard system existed for DC motors, however the major use for such lines was in precision controlled multimotor lines where synchronous AC motors were used for each section and when the master alternator frequency was varied, all the motors would follow together with synchronous accuracy. Such systems were still being installed on new machines as late as the mid-80s when static variable frequency controls became widely used. Static AC variable speed drives that were readily available were of the six step, variable voltage design. Later, when Phillips/ Signetics came out with a sine coded PWM chip set, sine coded PWM drives became the norm and six-step variable frequency faded into non-use except for unusual applications where the slightly lower loss at full speed, full load was an advantage.

5.3   CHARACTERISTICS OF AC DRIVES

The current source type is fairly common at 100 hp and higher and does offer some advantages. These two types will be discussed. One universal aspect of induction motor speed control is that slip is required in order for the motor to produce useful torque

[4]. The various means for improvement in performance such as voltage boost, energy savings by voltage reduction, slip compensation, "scalar" control, vector control, etc., do nothing to prevent the slip which is inherent in the motor design. At full speed, full load, an AC motor will slip an amount roughly equal to 3% of its synchronous speed based on the motor design, and this is not affected in the least by the design of the controller. This slip represents energy loss in the rotor of the AC motor and converts directly to heat inside the motor which must be dissipated. This one fact alone makes the AC machine relatively limited in terms of full torque thermal speed range. The standard induction motor cannot produce full torque over more than a 2 to 1 speed range without overheating if subjected to a continuous operation. Energy efficient types can stretch this speed range to approximately 3 to 1 and special designs can extend this to 10 to 1 [4, 3]. Dynamic response of the standard variable frequency drive is very limited with bandwidths of 1 to 3 Hz typical [7]. The PWM AC drive is the drive of choice for applications not requiring high starting torques, high continuous torque at low speeds, or fast dynamic response. Such applications are pumps, fans, some types of conveyors, and similar applications. It should be noted that by judicious use of a combination of high efficiency or special design motors, much higher torques can be achieved but at a higher cost. One of the most compelling reasons to use such a variable speed drive is price (in the small hp sizes particularly). Another reason why AC drives are nearly universally used in fan and pump applications is the ability to run a standard AC motor (if the application allows it) which can be bypassed, or operated from the AC line in the event the controller might become inoperative. The ability to do this in a municipal pumping application where sewage or water is pumped is imperative and in many cases, the locations where this equipment is located is remote and controlled by computer or PLC and the sensing of controller malfunction and bypassing it is all done automatically without human assistance. The problem is annunciated and is later fixed by maintenance personnel.

5.3.1   Applications of AC Drives or Variable Frequency Drives

Feature	Benefits
Soft starting	• Reduced impact on electrical network means no penalties from utility
	• Reduced stress on motor, coupling and load, giving extended life time
	• Unlimited number of starts per hour
Precise speed and torque control	• Better product quality
	• Improved cost of ownership
	• Better protection of motor (e.g., stall protection and load)
	• Consistent product quality, despite input power variations and sudden load changes
Wide speed control range	• Improved efficiency compared to traditional flow control methods, e.g., damper control, throttling
	• Lower maintenance

High reliability and availability	• Reduced downtime • Improved process availability
Low audible noise	• Improved working environment for operators
Capability for speed reversal/regenerative braking	• Desired torque during braking, therefore better product quality • Improved braking characteristics • Higher efficiency
Flux optimization (motor flux automatically adapted to load)	• Improved motor efficiency • Reduced motor noise
Power loss ride through	• Reduced number of drive trips • Better process availability
Automatic start (drive can catch a spinning load)	• Reduced waiting time • Reduced downtime
Energy saving	• AC drives can be retrofire to standard induction motors, to provide substantial energy savings

1. Fans
2. Blowers
3. Still mills
4. Cranes
5. Conveyors and
6. Traction etc.

5.3.2 Advantages and Applications of Induction Motor Drives

The main drawback of DC motors is the presence of commutate and brushes, which require frequent maintenance and make them unsuitable for explosive and dirty environments. On the other hand, induction motors, particularly squirrel cage is rugged, cheaper, lighter, smaller, and more efficient, require less maintenance and can operate in dirty and explosive environments.

Although variable speed induction motor drives are generally more expensive than DC drives, they are used in a number of applications such as fans, blowers, mill run-out tables, cranes, conveyors, traction, etc., because of the advantages of induction motors. Other applications involved are underground and underwater installations and explosive and dirty environments.

5.4 INDUCTION MACHINES

The stator winding for an induction machine looks like the stator winding of a synchronous machine. However, instead of an externally energized field winding, the induction machine has a rotor with, typically, copper or aluminium axial bars embedded in magnetic laminations at the outer periphery of the rotor. By energizing the stator windings with multiphase current, a rotating flux is created, just as in the synchronous motor. This flux induces currents in the rotor bars and creates a torque reaction. The rotor must rotate at an electrical speed lower or higher than that of the stator to produce the eddy currents and torque. If the rotor speed is lower, it will produce shaft torque to a load. If torque is applied to the shaft in the direction of rotation, the motor can operate at a rotor frequency higher than synchronous and deliver electrical power from the stator. The difference between shaft speed and synchronous speed is known as slip and is usually expressed as a per cent or per unit of synchronous speed. The induction motor can be operated as an induction generator with no external power if the excitation that provides the rotational flux is supplied from a capacitor bank on the stator. The machine will usually build up voltage self-excited from residual flux. However, it may be necessary to flash a voltage on the stator to start the process. The capacitance can be adjusted to operate the machine at rated voltage.

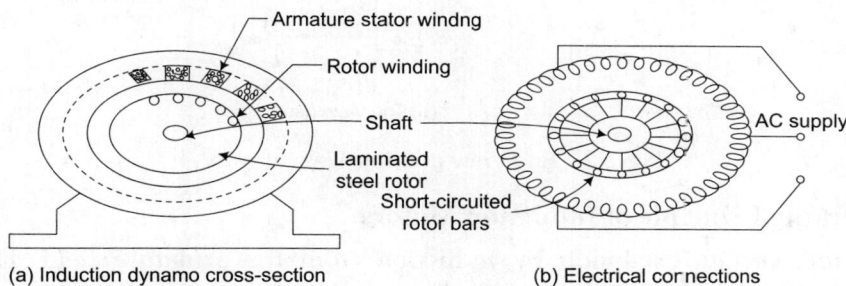

(a) Induction dynamo cross-section   (b) Electrical connections

Fig. 5.1   Construction of induction machine.

5.4.1   Construction of Induction Motor

The AC induction motor comprises two electromagnetic parts:
- Stationary part called the stator
- Rotating part called the rotor, supported at each end on bearings

The stator and the rotor are each made up of:
- An electric circuit usually made of insulated copper or aluminium, to carry current.
- A magnetic circuit usually made from laminated steel, to carry magnetic flux.

The stator is the outer stationary part of the motor, which consists of:

The outer cylindrical frame of the motor, which is made either of welded sheets steel, cast iron or cast aluminium alloy. This may include feet or a flange for mounting. The magnetic path, which comprises a set of slotted steel laminations pressed into the cylindrical space inside the outer frame. The magnetic path is laminated to reduce eddy currents, lower losses, and lower heating. The cross-sectional area of these windings must be large enough for the power rating of the motor. For a three-phase motor, 3 sets of windings are required, one for each phase.

This is the rotating part of the motor. As with the stator above, the rotor consists of a set of slotted steel laminations pressed together in the form of a cylindrical magnetic path and the electrical circuit. The electrical circuit of the rotor can be either: wound rotor type, which comprises 3 sets of insulated windings with connections brought out to 3 slip rings mounted on the shaft. The external connections to the rotating part are made via brushes onto the slip rings. Consequently, this type of motor is often referred to as a slip ring motor.

Squirrel cage rotor type, which comprises a set of copper or aluminium bars installed into the slots, which are connected to an end-ring at each end of the rotor. The construction of these rotor windings resembles a squirrel cage aluminium rotor bars are usually die-cast into the rotor slots, which results in a very rugged construction. Even though the aluminium rotor bars are in direct contact with the steel laminations, practically all the rotor current flows through the aluminium bars and not in the laminations.

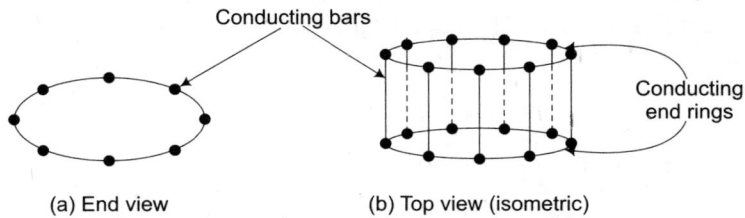

Conducting bars

Conducting end rings

(a) End view   (b) Top view (isometric)

Fig. 5.2   End view and top view of rotor.

5.4.2   Working Principle of Induction Motors

The rotor receives its excitation by induction from the armature field. Hence, the induction machine is a doubly-excited machine in the same sense as the synchronous and DC machines.

The basic principle of operation is described by Faraday's law. If we assume that the machine rotor is standstill, and the armature is excited, then the armature-produced rotating field is moving with respect to the rotor. In fact, the relative speed between the rotating field and the rotor is synchronous speed. For this condition, the rotating field induces a large voltage in the rotor bars. The large voltage causes a large current in the squirrel cage which, in turn, creates a magnetic field in the rotor. The rotor magnetic field interacts with the armature magnetic field and torque is produced. If the produced torque is larger than any load torque, the rotor begins to turn. As the rotor accelerates, the speed difference between the rotor and the armature field is reduced. This reduced speed difference (or slip) causes the induced rotor voltage to be reduced, the rotor current to be reduced, the rotor flux to be reduced, and the torque produced by the machine to be reduced. Eventually, the torque produced by the motor equals the torque demanded by the load, and the motor settles to an equilibrium rotor speed. This equilibrium rotor speed must be less than the synchronous speed since there must be a slip to produce torque.

5.4.3   Single-Phase Induction Motors

A single-phase induction motor in its simplest form is a polyphase induction motor having a squirrel cage rotor, the only difference is that single-phase induction motor has single winding on the stator which produces mmf stationary in space but alternating in time, a polyphase stator winding carrying balanced current, produces mmf rotating in space around the air gap and constant in time, with respect to an observer moving with the mmf. The stator winding of the single-phase motor is disposed in slots around the inner periphery of a laminated ring similar to a three-phase motor.

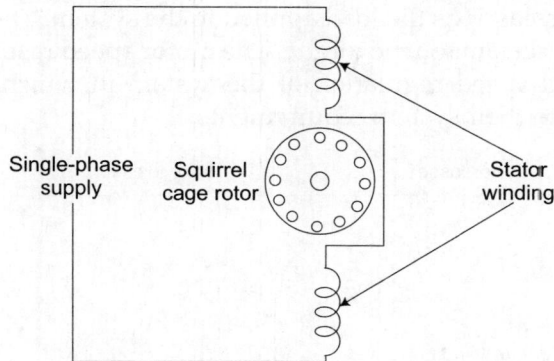

Fig. 5.3   Single-phase induction motor.

5.4.4   Split-Phase Induction Motor

The start winding is made with smaller gauge wire, and fewer turns relative to the main winding to create more resistance, thus putting the start winding's field at a different angle than that of the main winding, and causing the motor to rotate. The main winding, of heavier wire, keeps the motor running the rest of the time. A split-phase motor uses a switching mechanism that disconnects the start winding from the main winding when the motor comes up to about 75% of rated speed. In most cases, it is a centrifugal switch on the motor shaft. Switch mechanism disconnects start winding when the motor reaches three-fourths of rated speed.

5.4.5   Capacitor Start/Induction Run Induction Motor

It has a start-type capacitor in series with the auxiliary winding like the capacitor-start motor for high starting torque. And, like a PSC motor, it also has a run-type capacitor that is in series with the auxiliary winding after the start capacitor is switched out of the circuit. This allows high breakdown or overload torque.

Another advantage of the capacitor-start/capacitor-run type motor: It can be designed for lower full-load currents and higher efficiency.

The capacitor start motor also has a starting mechanism, either a mechanical or solid state electronic switch. This disconnects not only the start winding but also the capacitor when the motor reaches about 75% of rated speed. Capacitor start/induction run motors have several advantages over split-phase motors. Since the capacitor is in series with the start circuit, it creates more starting torque, typically 200 to 400% of rated load. And the starting current, usually 450 to 575% of rated current, is much lower than the split-phase due to the larger wire in the start circuit.

5.5   EQUIVALENT CIRCUIT OF INDUCTION MOTOR

The rotor current has an initial spike due to the initial voltage rush. As the load is applied to the motor, the current starts to oscillate between + 125 amps due to change in rotor flux. The stator flux increased as the load applied to the system. The reference torque has been followed by the electromagnetic torque. The motor speed results are not satisfactory because of the internal speed regulation of the system in which one cannot alter the parameters according to their system requirements.

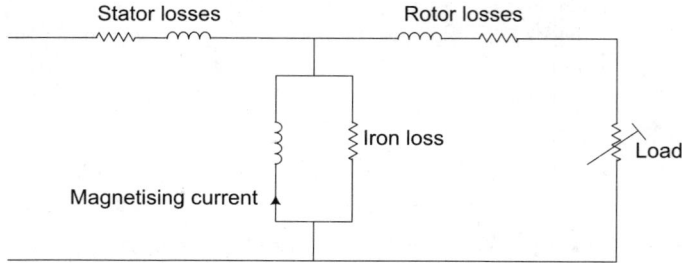

Fig. 5.4   Equivalent circuit of induction motor with stator loss and rotor loss.

Three-phase induction motors have either slip ring or cage rotors. They receive their excitation by magnetic induction from the stator side. These motors are quasi-constant speed motors. They have many applications in factories, workshops, large air-conditioning systems, large pumps, electric traction and large mechanical equipment.

The equivalent circuit of the three-phase induction motor is similar to that of the transformer with an added resistance representing the developed mechanical power.

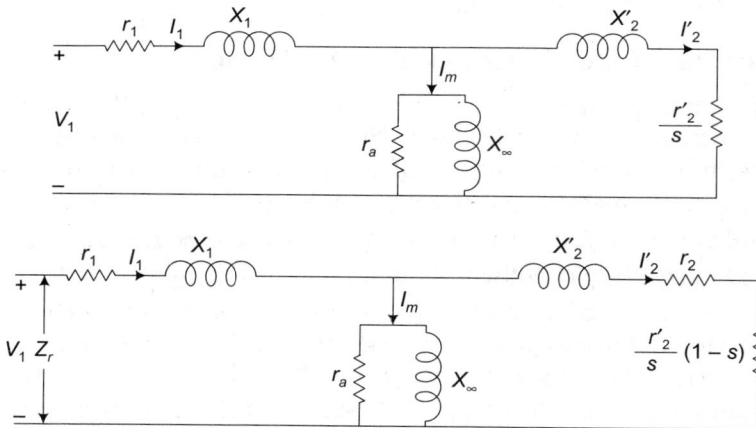

Fig. 5.5   Equivalent circuit of induction motor.

5.5.1   Mathematical Expression for rms Voltage

Frequency of rotation is given by:

$$\omega_s = \frac{2}{P} 2\pi f \text{ known as synchronous frequency}$$

where

P – Number of poles

f – Supply frequency

Rotating flux induced: Emf in stator winding (known as back emf)

Emf in rotor winding

Rotor flux rotating at synchronous frequency

Rotor current interact with flux to produce torque

Rotor always rotates at frequency less than synchronous, i.e., at slip speed

$$\omega_{sl} = \omega_s - \omega_r$$

Ratio between slip speed and synchronous speed is known as slip

$$s = \frac{\omega_s - \omega_r}{\omega_s}$$

Induced voltage

Flux density distribution in airgap: $B_{\max} \cos \theta$

Flux per pole:

$$\Phi_p = \int_{-\pi/2}^{\pi} (B_{\max} \cos \theta)\big| \, r \, d\theta$$

$$= 2 B_{\max} I_r$$

Sinusoidally distributed flux rotates at ω_s and induced voltage in the phase coils

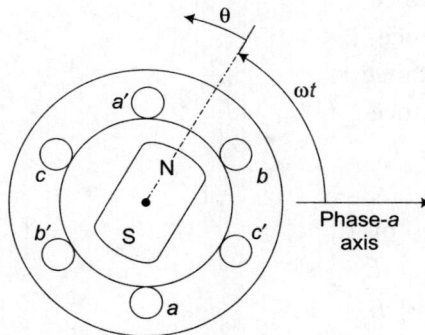

Fig. 5.6   Flux and speed of induction motor.

Maximum flux links phase a when $\omega t = 0$. No flux links phase a when $\omega t = 90°$

$$\lambda_a \equiv \text{flux linkage of phase a}$$

$$\lambda_a = N \Phi_p \cos (\omega t)$$

By Faraday's law, induced voltage in a phase coil aa is

$$e_a = -\frac{d\lambda}{dt} = \omega N \Phi_p \sin \omega t$$

$$E_{rms} = \frac{\omega N \Phi_p}{\sqrt{2}} = 4.44 f \, N\Phi_p$$

Maximum flux links phase a when $\omega t = 0$. No flux links phase a when $\omega t = 90°$

In actual machine with distributed and short-pitch windings, induced voltage is less than this by a winding factor K_w

$$E_{rms} = \frac{\omega N \Phi_p}{\sqrt{2}} = 4.44 f \, N\Phi_p K_w$$

$$E_{rms} = \frac{\omega N \Phi_p}{\sqrt{2}} = 4.44 f \, N\Phi_p$$

Stator phase voltage equation:

$$V_s = R_s I_s + j \, (2\pi f) \, L_{ls} I_s + E_{ag}$$

E_{ag} – air-gap voltage or back emf (E_{rms} derive previously)

$$E_{ag} = k f \phi_{ag}$$

Rotor phase voltage equation:

$$E_r = R_r I_r + js \, (2\pi f) \, L l_r$$

E_r – induced emf in rotor circuit

$$E_r/s = (R_r/s) \, I_r + j \, (2\pi f) L l_r$$

R_s – stator winding resistance
R_r – rotor winding resistance
L_{ls} – stator leakage inductance
L_{lr} – rotor leakage inductance
L_m – mutual inductance
s – slip

We know E_g and E_r related by

$$\frac{E_r}{E_{ag}} = \frac{s}{a}$$

where **a** is the winding turn ratio = N_1/N_2

The rotor parameters referred to stator are:

$$I_r' = \frac{I_r}{a}, \quad R_r' = a^2 R_r, \quad L_{lr}' = a^2 L_{lr}$$

$\therefore$ rotor voltage equation becomes

$$E_{ag} = (R_r'/s) \, I_r' + j \, (2\pi f) \, L_{lr}' I_r'$$

R_s –   stator winding resistance

R_r' –   rotor winding resistance referred to stator

L_{ls} –   stator leakage inductance

L_{lr}' –   rotor leakage inductance referred to stator

L_m –   mutual inductance

I_r' –   rotor current referred to stator

Power and torque expression: Power is transferred from stator to rotor via air–gap, known as air-gap power

$$P_{ag} = 3I_r'^2 \frac{R_r'}{s} = 3I_r'^2 R_r' + 3I_r'^2 \frac{R_r'}{s}[1-s]$$

$$3I_r'^2 R_r' = \text{Loss in rotor winding}$$

$$3I_r'^2 \frac{R_r'}{s}[1-s] = \text{Converted to mechanical power} = (1-s)\,P_{ag} = P_m$$

Mechanical power, $P_m = T_{em}\,\omega_r$

But, $s\omega_s = \omega_s - \omega_r \;\Rightarrow\; \omega_r = (1-s)\,\omega_s$

$\therefore$ $P_{ag} = T_{em}\,\omega_s$

Rotor current from equivalent circuit

$$I_r' = \frac{V_s}{\left(R_s + \dfrac{R_r'}{s}\right) + (X_{ls} + X_{lr}')}$$

$$T_{em} = \frac{P_{ag}}{\omega_s} = \frac{3I_r'^2 R_r'}{s\,\omega_s}$$

Therefore, torque is given by:

$$T_{em} = \frac{3R_r'}{s\,\omega_s} \cdot \frac{V_s^2}{\left(R_s + \dfrac{R_r'}{s}\right)^2 + (X_{ls} + X_{lr}')^2}$$

This torque expression is derived based on approximate equivalent circuit

A more accurate method is to use Thevenin equivalent circuit:

$$T_{em} = \frac{3R_r'}{s\,\omega_s} \cdot \frac{V_{Th}^2}{\left(R_{Th} + \dfrac{R_r'}{s}\right)^2 + (X_{Th} + X_{lr}')^2}$$

5.5.2   Calculation of the Parameters

DC TEST

$R_1 = V_{DC}/(2I_{DC})$ for Y connected stator

$R_1 = (3/2)\,V_{DC}/I_{DC}$ for delta connected stator

Blocked-rotor test

Let the three-phase input power to be P_{sc}, the input voltage to be V_{sc} and the input current to be I_{sc}. Referring to the motor equivalent circuit at zero speed (slip = 1) shown in **Fig. 5.9**.

$$Z_{eq} = V_{sc-p}/I_{sc-p}$$
$$R_{eq} = 1/3\, P_{sc}/I_{sc-p}^2$$
$$X_{eq} = [Z_{eq}^2 - R_{eq}^2]^{1/2}$$
$$R_2 = R_{eq} - R_1$$
$$X_1 = X_2 = X_{eq}/2$$

No-load test as a motor

Let the three-phase input power be P_o, the input line voltage be V_o and the input line current be I_o. Referring to the motor equivalent circuit at no-load shown in **Figs. 5.7 and 5.9**.

$$\Psi_o = \cos^{-1}[P_o/(\sqrt{3}\,V_o\,I_o)]$$
$$P_{rot} = P_o - 3I_{o-p}^2\,R_1$$

Then

$$\mathbf{E}_{o-p} = \mathbf{V}_{o-p} - \mathbf{I}_{o-p}\angle\Psi_o\,(R_1 + jX_1)$$
$$\cos\varphi_o = P_{rot}/(\sqrt{3}\,E_o I_o)$$
$$I_c = I_{o-p}\cos\varphi_o$$
$$R_c = E_{o-p}/I_c$$
$$I_m = I_{o-p}\sin\varphi_o$$
$$X_m = E_{o-p}/I_m$$

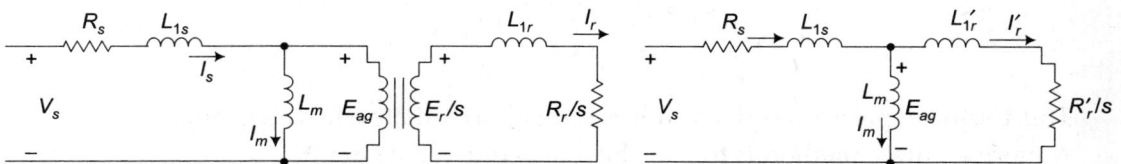

Fig. 5.7   Exact equivalent circuits with referred parameter.

Starting Methods of Three-Phase Induction Motors

As we know, once a supply is connected to a three-phase induction motor a rotating magnetic field will be set up in the stator, this will link and cut the rotor bars which, in turn, will induce rotor currents and create a rotor field which will interact with the stator field and produce rotation. Of course, this means that the three-phase induction motor is entirely capable of self-starting. The need for a starter, therefore, is not, conversely enough, to provide starting but to reduce heavy starting currents and provide overload and no-voltage protection.

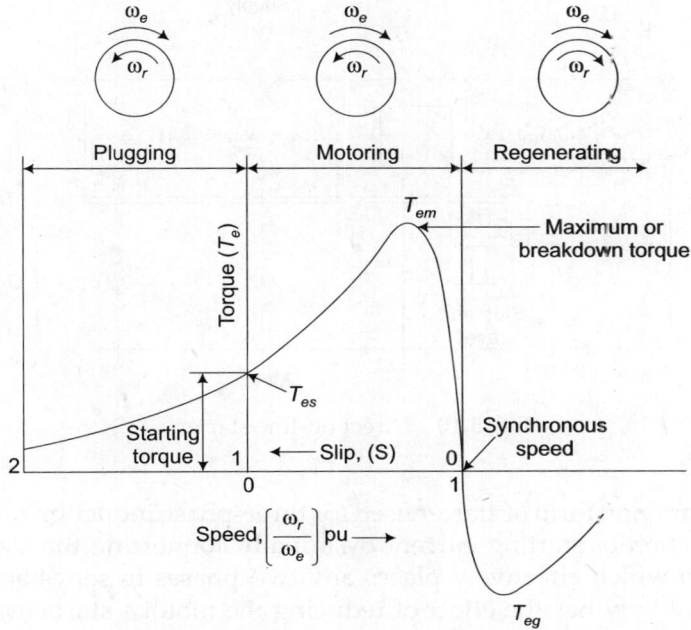

Fig. 5.8   Torque-speed characteristics of three-phase induction motor.

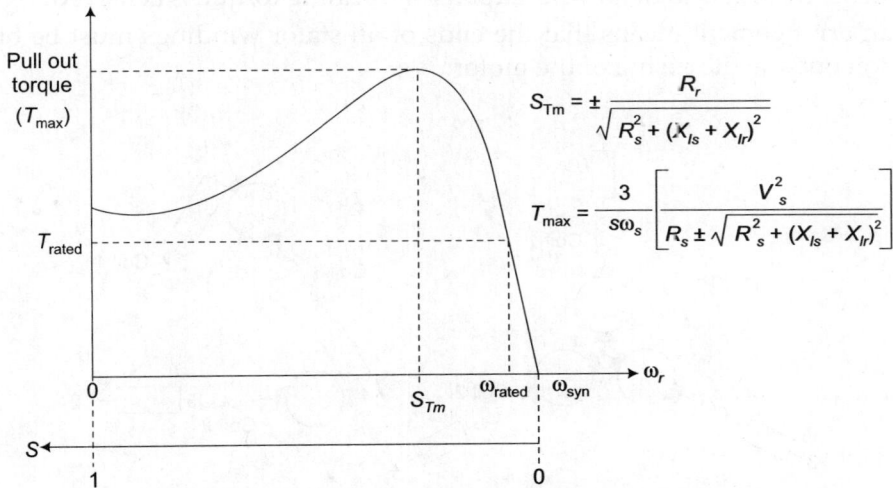

$$S_{Tm} = \pm \frac{R_r}{\sqrt{R_s^2 + (X_{ls} + X_{lr})^2}}$$

$$T_{max} = \frac{3}{s\omega_s}\left[\frac{V_s^2}{R_s \pm \sqrt{R_s^2 + (X_{ls} + X_{lr})^2}}\right]$$

Fig. 5.9   Torque-speed characteristics.

There are a number of different types of starter including 'the Direct On-line Starter', 'the Star-Delta Starter', 'Auto-Transformer' and 'Rotor resistance'.

Direct-on-Line Starter (DOL)

The DOL starter switches the supply directly on to the contacts of the motor. As the starting current of an induction motor can be 6-8 times the running current the DOL starter is typically only used for motors with a rating of less than 5 kW.

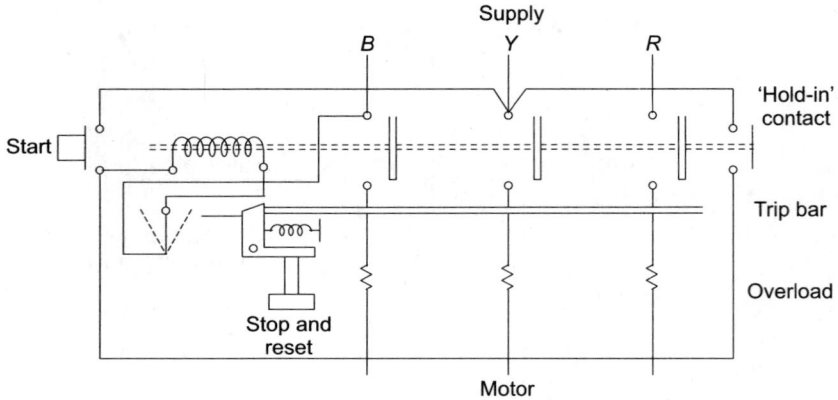

Fig. 5.10   Direct on-line starter.

Star Delta Starter

This is the most common form of starter used for three-phase induction motors. It achieves an effective reduction of starting current by initially connecting the stator windings in star configuration which effectively places any two phases in series across the supply. Starting in star not only has the effect of reducing the motor's start current but also the starting torque.

Once up to a particular running speed a double throw switch changes the winding arrangements from star to delta whereupon full running torque is achieved.

Such an arrangement means that the ends of all stator windings must be brought to termination outside the casing of the motor.

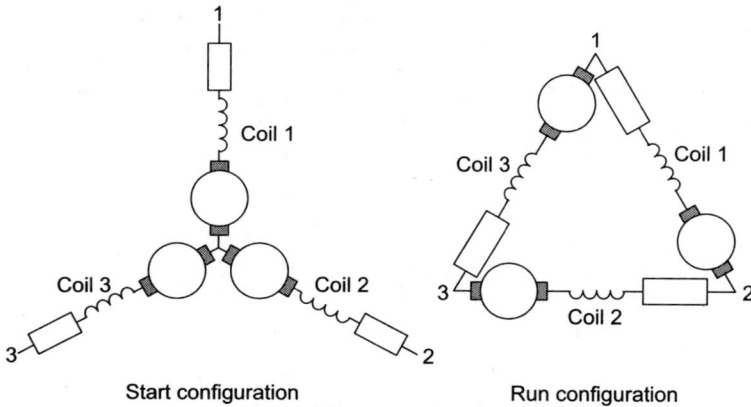

Fig. 5.11   Star-delta starter.

Auto-Transformer Starting

This method of starting reduces the start current by reducing the voltage at start-up. It can give lower start-up currents than star-delta arrangements but with an associated loss of torque.

It is not as commonly utilized as other starting methods but does have the advantage that only three connection conductors are required between the starter and motor.

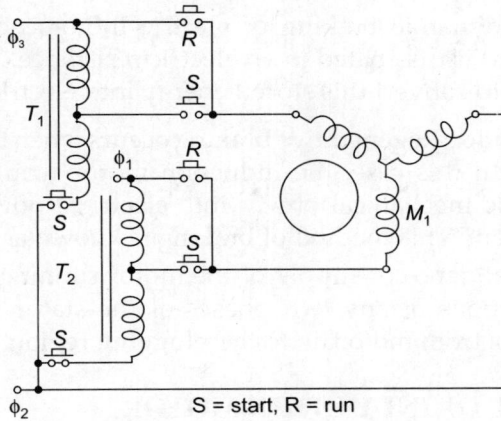

Fig. 5.12 Autotransfer starter.

Rotor Resistance Starter

If it is necessary to start a three-phase induction motor on load then, a wound rotor machine will normally be selected. Such a machine allows an external resistance to be connected to the rotor of the machine through slip rings and brushes.

At start-up, the rotor resistance is set at maximum but is reduced as speed increases until eventually it is reduced to zero and the machine runs as if it is a cage rotor machine.

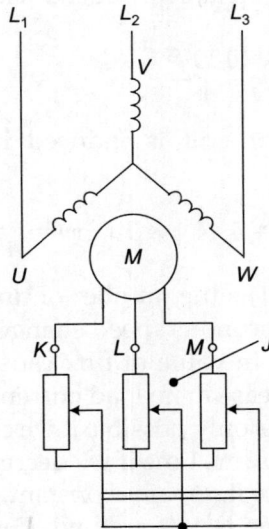

Fig. 5.13 Rotor resistance starter.

5.5.3   Braking Methods of Induction Motor Drive

There are three modes of braking in an induction motor.

Generating mode: The dynamic braking of electric motors occurs when the energy stored in the rotating mass is dissipated in an electric resistance. This requires the motor to operate as a generator to convert this stored energy into electrical.

Regenerative braking mode: Regenerative braking occurs when the motor speed exceeds the synchronous speed. In this case, the induction motor would run as an induction machine is converting the mechanical power into electrical power, which is delivered back to the electrical system. This method of braking is known as regenerative braking.

Plugging: When phase sequence of supply of the motor running at a speed is reversed, by interchanging connections of any two phases of the stator with respect to supply terminals, operation shifts from motoring to the plugging region.

5.6   SPEED CONTROL OF INDUCTION MOTOR

There are four methods of speed control of induction motor:
1. Stator voltage control
2. Supply frequency control
3. Rotor resistance control
4. Slip power recovery control

Speed Control from Stator Side

V/f control or frequency control: Whenever three-phase supply is given to three-phase induction motor rotating magnetic field is produced which rotates at synchronous speed given by:

$$N_s = \frac{120 f}{P}$$

In three-phase induction motor emf is induced by induction similar to that of transformer which is given by:

$$E \text{ or } V = 4.44 \, \phi \, K.T.f \text{ or } f = \frac{V}{4.44 \, KTf}$$

where K is the winding constant, T is the number of turns per phase and f is frequency. Now if we change frequency synchronous speed changes but with decrease in frequency flux will increase and this change in value of flux causes saturation of rotor and stator cores which will further cause increase in no load current of the motor . So, it is important to maintain flux , ϕ constant and it is only possible if we change voltage, i.e., if we decrease frequency flux increases but at the same time if we decrease voltage flux will also decease causing no change in flux and hence it remains constant. So, here we are keeping the ratio of V/f as constant. Hence, its name is V/f method. For controlling the speed of three-phase induction motor by V/f method, we have to supply variable voltage and frequency which is easily obtained by using converter and inverter set.

Controlling supply voltage: The torque produced by running three-phase induction motor is given by:

$$T \propto \frac{s\,E_2\,R_2}{R_2^2\ (sX_2)^2}$$

In low slip region $(sX)^2$ is very small as compared to R_2. So, it can be neglected. So torque becomes

$$T \propto \frac{s\,E_2^2}{R_2}$$

Since rotor resistance, R_2 is constant, so the equation of torque further reduces to

$$T \propto sE_2^2$$

We know that rotor induced emf $E_2 \propto V$. So, $T \propto sV^2$. From the equation above it is clear that if we decrease supply voltage torque will also decrease. But for supplying the same load, the torque must remain the same, and it is only possible if we increase the slip and if the slip increases the motor will run at reduced speed. This method of speed control is rarely used because a small change in speed requires a large reduction in voltage, and hence the current drawn by motor increases, which causes overheating of induction motor.

Changing the number of stator poles

The stator poles can be changed by two methods:

• Multiple stators winding method
• Pole amplitude modulation method (PAM)

Multiple stators winding method: In this method of speed control of three-phase induction motor, the stator is provided by two separate windings. These two stator windings are electrically isolated from each other and are wound for two different pole numbers. Using switching arrangement, at a time, supply is given to one winding only and hence speed control is possible. Disadvantages of this method are that smooth speed control is not possible. This method is more costly and less efficient as two different stator windings are required. This method of speed control can only be applied to squirrel cage motor.

Pole amplitude modulation method (PAM): In this method of speed control of three-phase induction motor the original sinusoidal mmf wave is modulated by another sinusoidal mmf wave having different number of poles.

Let $f_1(\theta)$ be the original mmf wave of induction motor whose speed is to be controlled
$f_2(\theta)$ be the modulation mmf wave
P_1 be the number of poles of induction motor whose speed is to be controlled
P_2 be the number of poles of modulation wave

$$f_1(\theta) = F_1 \sin\left(\frac{P_1\,\theta}{2}\right)$$

$$f_2(\theta) = F_2 \sin\left(\frac{P_2\,\theta}{2}\right)$$

After modulation-resultant mmf wave

$$F_r(\theta) = F_1 F_2 \sin\left(\frac{P_1 \theta}{2}\right) \sin\left(\frac{P_2 \theta}{2}\right)$$

Apply formulae for $2 \sin A \sin B = \cos\dfrac{A-B}{2} - \cos\dfrac{A+B}{2}$

So we get, resultant mmf wave

$$F_r(\theta) = F_1 F_2 \frac{\cos\dfrac{(P_1 - P_2)\theta}{2} - \cos\dfrac{(P_1 + P_2)\theta}{2}}{2}$$

Therefore, the resultant mmf wave will have two different numbers of poles i.e.,

$$P_{11} = P_1 - P_2 \quad \text{and} \quad P_{12} = P_1 + P_2$$

Therefore, by changing the number of poles we can easily change the speed of three-phase induction motor.

Adding rheostat in the stator circuit: In this method of speed control of three-phase induction, motor rheostat is added in the stator circuit due to this voltage drops. In the case of three-phase induction motor torque produced is given by $T \propto sV_2^2$. If we decrease supply voltage torque will also decrease. But for supplying the same load, the torque must remain the same, and it is only possible if we increase the slip and if the slip increased motor will run at reduced speed.

Supplying the stator a variable voltage, variable frequency supply using static frequency converters can control the induction motor speed. Speed control is also possible by feeding the slip power to the supply system using converters in the rotor circuit; basically one distinguishes two different methods of speed control.

(a) Speed control by varying the slip frequency when the stator is fed from a constant voltage, constant frequency mains.

(b) Speed control of the motor using a variable frequency variable voltage motor operating at constant rotor frequency.

5.6.1   Control of Induction Machines in the Dynamic State

Various modulation techniques are used to study the transient performance of the machine under dynamic conditions. A generally used one is a dynamic d-q model, space vector modulation, and spiral vector modulation. Out of these, the dynamic d-q model is discussed here in detail.

Dynamic d-q model

Vector control techniques for AC motor drive systems have gained wide acceptance in high-performance, variable-speed applications by creating independent channels for torque and flux controls. In a similar manner, vector control strategies have been proposed for active and reactive power control of the induction generator. Stator and rotor currents flowing through balanced sinusoidal distributed windings set up respective resultant space mmf vectors which may be defined in terms of the current space vectors I_s and I_r

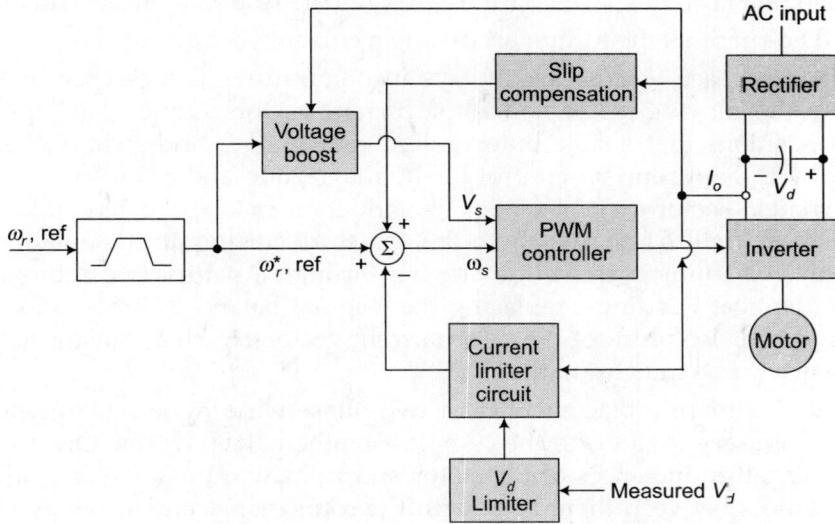

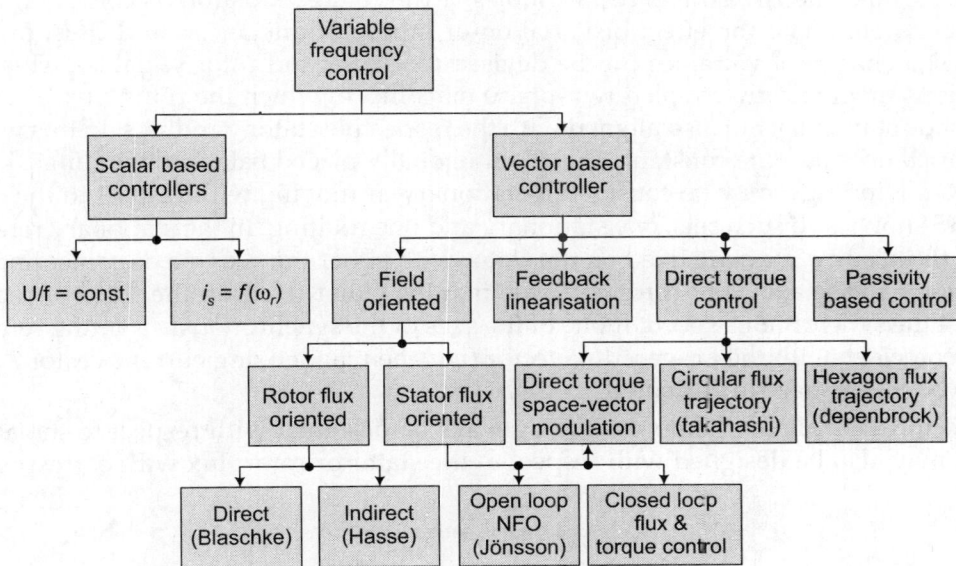

Fig. 5.14 Speed controls of induction motor drives.

as shown in **Fig. 5.14**. The developed electromagnetic torque is proportional to the product of the magnitudes of the two current vectors and the sine of their space phase difference, i.e.,

$$T_e = K I_s I_r \sin \theta = K I_s I_m \sin \delta_s = K I_r I_m \sin \delta_r$$

I_m is the magnetizing current space vector, represents the resultant air-gap flux vector $I_s \sin \delta_s$ ($I_r \sin \delta_r$) represents the torque producing a current vector.

These two rotating space vectors are always in quadrature. The essence of vector control is to force the moving stator and rotor current vectors I_s and I_r to take these magnitudes and positions that enable independent control of I_m and $I_s \sin \delta_s$ ($I_r \sin \delta_r$). This is achieved by the appropriate control of the magnitude and phase of the actual stator (rotor) currents. Vector control makes an induction machine behave like a DC machine with $I_s \sin \delta_s$ ($I_r \sin \delta_r$) analogous to the armature current and I_m analogous to the field excitation. Assuming currents are flowing through a pair of two orthogonally spaced fictitious identical windings, replacing the original balanced three-phase stator and rotor windings can also produce the same current vectors. Such a transformation is known as *a* reference frame transformation.

However, for this, a mere replacement by a two-phase winding is not sufficient. A further insight is necessary to develop the complete mathematical model. Owing to the smooth air gap, the self-inductances of the stator and rotor windings are constant, but the mutual inductances between them vary with the rotor displacement relative to the stator. This variation of the stator-to-rotor mutual inductances makes the induction motor analysis complicated in terms of real variables, as the voltage equations become non-linear. In order to eliminate the effect of variation of mutual inductances, and thus, facilitate analysis, a change of variables can be devised for stator and rotor variables. This gives a fictitious magnetically coupled two-phase machine, in which the rotor circuits are not only made stationary but also aligned with the respective stator windings. In this way, all the inductances become constant. These orthogonally placed balanced windings, known as the *d-q* windings, may be considered stationary or moving with respect to the stator. **Fig. 5.15** shows two such sets, one stationary and one rotating. In the stationary reference frame, the d^s and q^s axes are fixed on the stator with either the d^s or q^s axis coinciding with the stator a-phase axis. The rotor d^e-q^e axis may be either fixed on the rotor or made to move at the synchronous speed. If one of the axes of the synchronously rotating reference frame coincides with the air-gap flux vector (i.e., the magnetizing current vector I_m), it is said to be the air-gap flux oriented.

A vector-controlled scheme need not always be designed with respect to the air-gap flux. It may also be designed with respect to the stator or rotor flux with corresponding

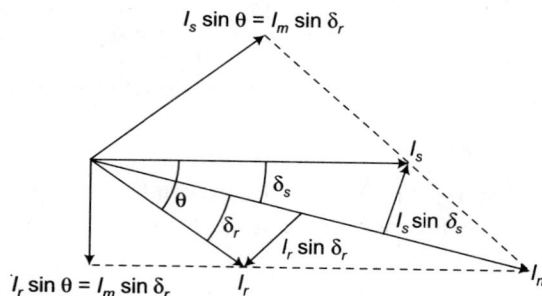

Fig. 5.15 Spatial mmf phasor diagram.

advantages or limitations. In field oriented control (FOC), the stator phase currents are first estimated in a synchronously rotating reference frame and then transformed back to the stationary stator frame to feed the machine. To carry out the transformation, with the invariance of power as the necessary criterion, and assuming the equivalent two-phase windings to have √3/2 times as many turns per phase as the three-phase winding, the fictitious stator d, q, 0 variables are obtained from the state variables (a, b, c) through a transform (similar to the park transform in synchronous machines) defined as:

$$
\begin{bmatrix} f^e_{ds} \\ f^e_{qs} \\ f^e_{os} \end{bmatrix} = \sqrt{\frac{2}{3}} \begin{bmatrix} \cos\theta_e & \cos\left(\vartheta_e - \dfrac{2\pi}{3}\right) & \cos\left(\theta_e + \dfrac{2\pi}{3}\right) \\ -\sin\vartheta_e & -\sin\left(\theta_e - \dfrac{2\pi}{3}\right) & -\sin\left(\theta_e + \dfrac{2\pi}{3}\right) \\ \dfrac{1}{\sqrt{2}} & \dfrac{1}{\sqrt{2}} & \dfrac{1}{\sqrt{2}} \end{bmatrix} \times \begin{bmatrix} f_{as} \\ f_{bs} \\ f_{cs} \end{bmatrix}
$$

where θ_e is the angle of the moving d^e axis with respect to the stator a-phase winding as shown in **Fig. 5.16.**

In the equation above, f can represent voltage, current or flux-linkage. This transformation is based on the assumption of a distributed sinusoidal winding. The phase variables are obtained from the d, q, 0 variables through the inverse of the transformation matrix in the above equation.

$$
\begin{bmatrix} f_{as} \\ f_{bs} \\ f_{cs} \end{bmatrix} = \sqrt{\frac{2}{3}} \begin{bmatrix} \cos\theta_e & -\sin\vartheta & \dfrac{1}{\sqrt{2}} \\ \cos\left(\vartheta_e - \dfrac{2\pi}{3}\right) & -\sin\left(\theta_e - \dfrac{2\pi}{3}\right) & \dfrac{1}{\sqrt{2}} \\ \cos\left(\theta_e + \dfrac{2\pi}{3}\right) & -\sin\left(\theta_e + \dfrac{2\pi}{3}\right) & \dfrac{1}{\sqrt{2}} \end{bmatrix} \times \begin{bmatrix} f^e_{ds} \\ f^e_{qs} \\ f^e_{os} \end{bmatrix}
$$

With reference to **Fig. 5.16** replacing θ_e by ($\theta_e - \theta_r$) in the above equations, defines the same transformations for the rotor quantities. The stator d, q variables in the reference

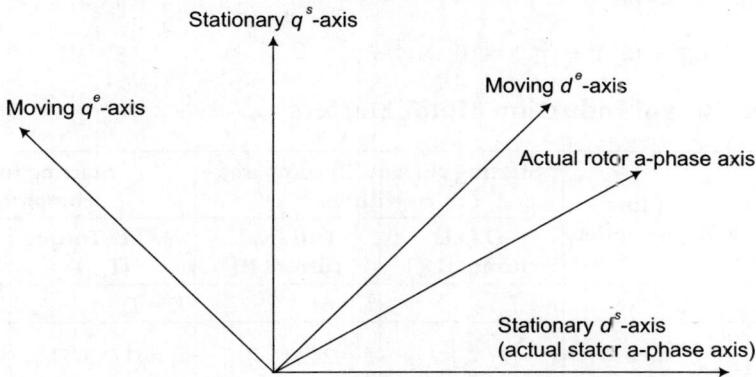

Fig. 5.16   Angular relationships between reference axis.

frame fixed to the stator, with the d-axis aligned along the a-phase axis, are related to the phase variables as follows:

$$
\begin{bmatrix} f^e_{ds} \\ f^e_{qs} \\ f^e_{os} \end{bmatrix} = \sqrt{\frac{2}{3}} \begin{vmatrix} 1 & -1/2 & -1/2 \\ 0 & \sqrt{3}/2 & -\sqrt{3}/2 \\ \dfrac{1}{\sqrt{2}} & \dfrac{1}{\sqrt{2}} & \dfrac{1}{\sqrt{2}} \end{vmatrix} \times \begin{bmatrix} f_{as} \\ f_{bs} \\ f_{cs} \end{bmatrix}
$$

In the synchronously rotating reference frame defined by the $d^e - q^e$ axes, the dynamic voltage equations of a three-phase symmetrical induction machine in terms of the equivalent two-phase system, defined by the previous equations are given by:

$$v^e_{ds} = r_s \, i^e_{ds} + p\, \lambda^e_{ds} - \omega_e\, \lambda^e_{qs}$$

$$v^e_{qs} = r_s \, i^e_{qs} + p\, \lambda^e_{qs} + \omega_e\, \lambda^e_{ds}$$

$$v^e_{dr} = r'_r \, i^e_{dr} + p\, \lambda^e_{dr} - (\omega_e - \omega_r)\, \lambda^e_{qr}$$

$$v^e_{qr} = r'_r \, i^e_{qr} + p\, \lambda^e_{qr} + (\omega_e - \omega_r)\, \lambda^e_{dr}$$

The electromagnetic torque in terms of the rotor currents is:

$$T_e = M\,(p/2)(i^e_{qs} i^e_{dr} - i^e_{ds} i^e_{qr})$$

where M is the mutual inductance and p is the number of poles.

Active power $\qquad\qquad P = (v^e_{ds} i^e_{ds} + v^e_{qs} i^e_{qs})$

Reactive power $\qquad\quad Q = (v^e_{qs} i^e_{ds} - v^e_{ds} i^e_{qs})$

and

$$(v^e_{ds})^2 + (v^e_{qs})^2 = 3\,V^2$$

where V is the rms input voltage. For balanced sets, v_0 and i_0 will be zero. Whether balanced or not, the relations given below always hold good:

$$P = (v^e_{ds} i^e_{ds} + v^e_{qs} i^e_{qs} + v^e_{0s} i^e_{0s}) = v_a i_a + v_b i_b + v_c i_c$$

and

$$(v^e_{ds})^2 + (v^e_{qs})^2 + (v^e_{0s}) = v^2_{as} + v^2_{bs} + v^2_{cs}$$

5.6.2   Comparison of Induction Motor Starters

Description of starter	% of line voltage applied	Starting current (I_s) compared with		Starting torque (T_s) compared with	
		D.O.L current (I_{dol})	Full load current (I)	D.O.L. Torque (T_{dol})	Full load torque (T)
D.O.L Starter	100%	$I_s = I_{dol}$	$I_s = 6I$	$T_s = T_{dol}$	$T_s = 6\,T$
Star delta starter	57.7%	$I_s = (1/\sqrt{3})^2\, I_{dol}$	$I_s = 2I$	$T_s = (1/\sqrt{3})^2\, T_{dol}$	$T_s = 2/3\,T$

Auto transformer starter	80%	$I_s = (0.8)^2 I_{dol}$	$I_s = 3.84\ I$	$T_s = (0.8)^2 T_{dol}$	$T_s = 1.28\ T$
	60%	$I_s = (0.6)^2 I_{dol}$	$I_s = 2.16\ I$	$T_s = (0.6)^2 T_{dol}$	$T_s = 0.72\ T$
	40%	$I_s = (0.4)^2 I_{dol}$	$I_s = 0.96\ I$	$T_s = (0.4)^2 T_{dol}$	$T_s = 0.32\ T$
Reactance-resistance starter	64%	$I_s = (0.64)^2 I_{dol}$	$I_s = 2.5\ I$	$T_s = (0.425)^2 T_{dol}$	$T_s = 0.35\ T$

5.7 STATIC SLIP RECOVERY OF INDUCTION MOTOR

With the production of silicon diode and silicon controlled rectifier of adequately large ratings, it is possible to replace all auxiliary machines of the traditional Scherbius drive by a three-phase bridge rectifier and static inverter resulting in more compact static slip energy recovery (SER) controlled slip ring induction motor drive. The static SER induction motor drive is popular among variable speed AC drives because of its inherent characteristics of higher efficiency, low converter cost, and a simple control circuit. It is being used in industrial applications such as crane, hoist, pump, and fan. The speed control of the drive is achieved by regulating the rotor voltage through the firing angle of the phase commutated inverter. When the firing angle is increased, the high DC output voltage of the inverter prohibits the flow of rotor current, resulting in low generation of torque, thereby, low speed of the drive. On the other hand, if the firing angle is increased, small DC output voltage of the inverter permits the flow of higher rotor current, resulting in high torque and high speed. They included the motor leakage reactance in steady-state analysis. Presented the starting transients of static SER induction motor drive operating in sub-synchronous range of speed using a set of non-linear differential equations developed using the d-q model in synchronously rotating reference frame. The non-linear differential equations were simulated on a digital computer and solved using Runga-Kutta method.

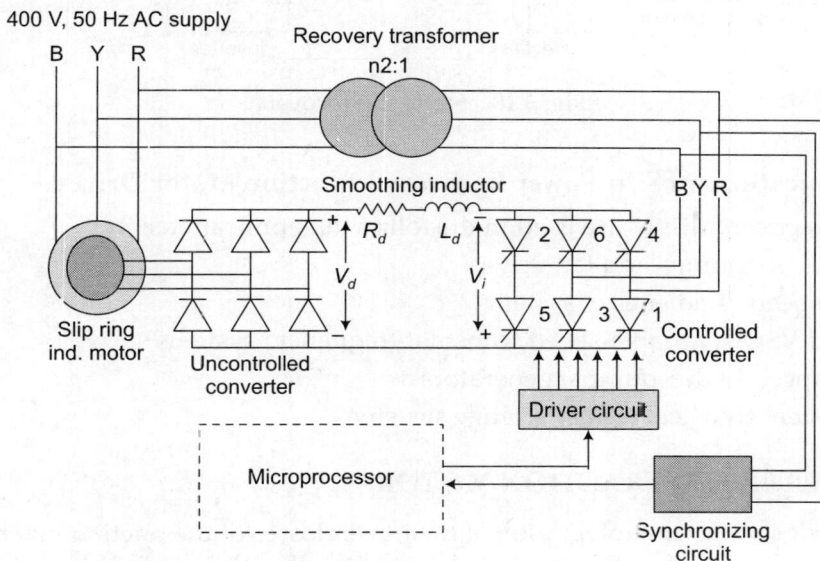

Fig. 5.17 Static slip recovery of induction motor.

The effects of firing angle, load, system inertia and filter time constant on transient torques and speeds following a switching operation were investigated. Brown et al. [6] proposed a model which was a hybrid between the traditional d-q and a – b – c models in which the rotor circuit was kept intact with transformation applied to the stator circuits only. Using the proposed model, the steady-state behaviour of the drive was predicted. Corrected the assumptions of the model used in the paper [5] and predicted the steady-state performance of the drive operating in subsynchronous range of speed using synchronously rotating reference frame model. Developed a hybrid model of the drive in which actual rotor phase variables were retained and transformed the stator phase variables. The steady state and transient performances of the drive were examined considering commutation overlap and harmonics of the rectifier and the inverter. The inclusion of effects of commutation overlap and harmonics added the complexity of the model. A developed simplified 5th order d-q dynamic model utilizing the arbitrary reference frame and predicted the dynamic behaviour of SER controlled slip ring induction motor drive.

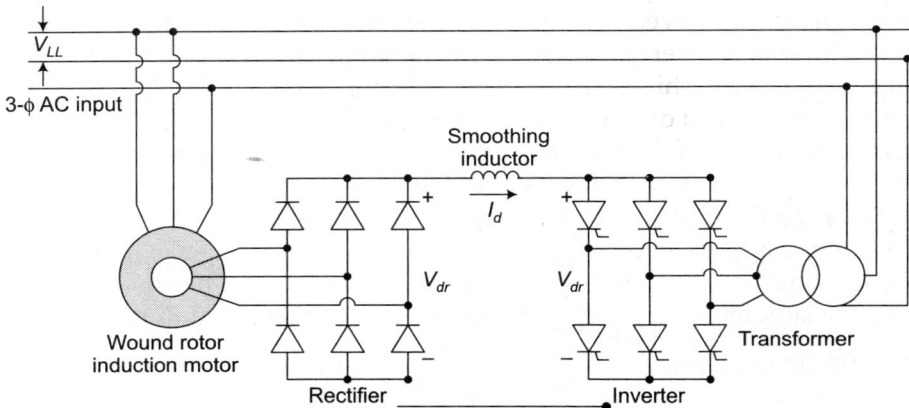

Fig. 5.18   Static slip recovery.

5.7.1   Applications of Slip-Power Recovery Induction Motor Drives

Slip-power recovery drives are used in the following applications:
• Large-capacity pumps and fan drives
• Variable-speed wind energy systems
• Shipboard VSCF (variable-speed/constant frequency) systems
• Variable speed hydro-pumps/generators
• Utility system flywheel energy storage systems

5.8   INVERTER FED TRACTION MOTOR

This section deals with the integration of the power losses of the traction inverter system. The inverter is assumed to be a standard six switch three-phase bridge and is assumed to use a switching frequency of 10 kHz. The main aim is to provide a suitable means for system-level simulation, in order to calculate the conduction and switching losses of the power devices within an inverter drive in an automotive propulsion environment. The efficiency of the inverter can then be easily determined from the modelled losses. A typical three-phase inverter drive is used for propulsion applications. An averaging technique is used to model the inverter component-level switching losses. The models are developed such that they are compatible.

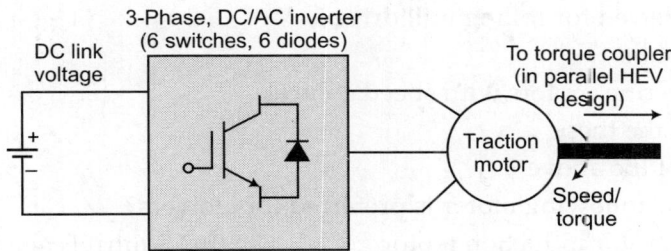

Fig. 5.19   Inverter fed reaction motor.

Multiple-Choice Questions

1. Stator voltage control for speed control of induction motors is suitable for:
   - (a) Drive of a crane
   - (b) Fan and pump drives
   - (c) Running it as generator
   - (d) Constant load drive
2. A wound rotor induction motor is preferred over squirrel cage induction motor when the major consideration involved is:
   - (a) Higher starting torque
   - (b) Low starting current
   - (c) Speed control over limited range
   - (d) All of the above
3. The motor normally used for crane travel is:
   - (a) Ward Leonard controlled DC shunt motor
   - (b) Synchronous motor
   - (c) AC slip ring motor
   - (d) DC differentially compound motor
4. In motor circuit, static frequency changers are used for
   - (a) Power factor improvement
   - (b) Improved cooling
   - (c) Speed regulation
   - (d) Reversal of direction
5. A pole changing type squirrel cage motor used in derricks has four, eight and twenty-four poles. In this, the medium speed is used for
   - (a) Lifting
   - (b) Hoisting
   - (c) Lowering
   - (d) Landing the load

6. A pole changing type squirrel cage motor used in derricks has four, eight and twenty-four poles. In this, the lowest speed is used for
   (a) Lifting   (b) Landing the load
   (c) Hoisting   (d) Lowering
7. The range of horsepower of electric motor drives for rolling mills is of the order of
   (a) 1 to 10 hp   (b) 15 to 25 hp
   (c) 50 to 100 hp   (d) 100 to 500 hp
8. Motors preferred for rolling mill drive is
   (a) DC motors
   (b) AC slip ring motors with speed control
   (c) Any of the these
   (d) None of the above
9. Which of the following motor is preferred for blowers?
   (a) Wound rotor induction motor   (b) Squirrel cage induction motor
   (c) DC shunt motor   (d) DC series motor
10. Centrifugal pumps are usually driven by
   (a) DC shunt motors   (b) Squirrel cage induction motors
   (c) DC series motors   (d) Any of the these

Answer

1. (b)	2. (d)	3. (c)	4. (c)	5. (a)
6. (b)	7. (d)	8. (c)	9. (b)	10. (b)

Exercise

1. A three-phase squirrel cage induction motor drives a blower type load. No-load rotational losses are negligible. Show that the rotor current is maximum when the motor runs at a slip equal to 1/3. Find the expression for maximum rotor current.
2. Explain the mechanical characteristics of a three-phase induction motor with stator current control.
3. Discuss how the speed of a three-phase induction motor can be controlled by varying the frequency of the applied voltage.
4. With necessary diagram, explain the theoretical principles of stator voltage control.
5. Bring out the limitations of stator voltage control scheme.
6. Derive an expression for the torque of an inverter fed three-phase induction motor when it is operated with V/f control. Show that the maximum torque remains unaltered in this scheme.

7. Explain in detail the speed control scheme for a three-phase induction motor using PWM inverter. Provisions are to be made in the scheme for speed control as well as regenerative braking in both directions.

8. Explain the operation of V/f control technique of speed control method of the induction motor. List the ways to implement the voltage to frequency ratio.

9. Explain using a power circuit how the speed of a diode bridge based voltage source inverter fed from induction motor drive can be controlled.

10. Explain using a power circuit how the speed of an induction motor drive can be controlled by using current source inverter.

11. Bring out the advantages of CSI over VSI fed induction motor drives.

12. Develop an expression relating speed and torque of a three-phase induction motor, whose speed is controlled using static rotor resistance control scheme.

13. Explain static rotor resistance control in closed loop speed control.

14. Explain using a power circuit the working of a static Kramer-drive system.

15. Show that the no-load speed of the induction motor in the Kramer drive can be varied from near standstill to full speed as the firing angle α is varied from almost 180 degrees to 90 degrees.

16. Explain using a diagram the working of a static Scherbius system. Show that it can operate in the synchronous, sub-synchronous and super-synchronous ranges. Dicuss its advantages.

17. Discuss the methods of improving the power factor of the sub-synchronous static converter cascade.

18. Explain closed loop speed control schemes for CSI and VSI fed drives.

19. Explain how can be control of induction machines in the dynamic state.

20. Write an application of slip-power recovery induction motor drives.

Numerical Problems

1. A 400 V, 4-pole, 50 Hz, three-phase star connected induction motor has $r_1 = 0$, $x_1 = x_2 = 1\ \Omega$, $r_2 = 0.4\ \Omega$, $x_m = 500\ \Omega$. The induction motor is fed from
   (a) A constant voltage source of 231 volts per phase and
   (b) A constant current source of 28 A. For both the cases calculate the slip at which maximum torque occurs and the starting and maximum torques.

2. A three-phase 440 V, 50 Hz, 6-pole star connected induction motor has the following parameters referred the stator $r_1 = 0.5\ \Omega$, $x_1 = x_2 = 1\ \Omega$, $r_2 = 0.6\ \Omega$. The stator to rotor turns ratio is 2. If the motor is regenerative braked, determine:
   (a) The maximum overhauling torque, it can hold and the range of speed in which it can safely operate.
   (b) The speed at which it will hold a load with a torque of 160 Nm.

3. A three-phase 400 V, 15 kW, 1440 rpm 50 Hz star connected induction motor has rotor leakage impedance of $(0.4 + j\,1.6)$ Ω. The stator leakage impedance and rotational losses are assumed negligible. If the motor is energized from a 120 Hz, 400 V, three-phase source, calculate:
   (a) Motor speed at rated load
   (b) The slip at maximum torque occurs
   (c) The maximum torque

4. A three-phase star connected 60 Hz, 4-pole induction motor has the following parameters for equivalent circuit $R_s = R'_r = 0.024$ Ω and $X_s = X'_r = 0.12$ Ω. The motor is controlled by variable frequency control with constant (V/f) ratio for operating frequency of 12 Hz. Calculate:
   (a) The breakdown torque as a ratio of its value at the rated frequency for both motoring and braking.
   (b) The starting torque and rotor current in terms of their values at the rated frequency.

5. A 100 hp, 460 V, 60 Hz star connected squirrel cage induction motor has the eq.ckt parameters $r_1 = 0.06$ Ω, $x_1 + x'_2 = 0.6$ Ω, $r'_2 = 0.35$ Ω, $X_m = 8$ Ω. The motor drives a fan which requires 100 hp at a speed of 1000 rpm. Determine the firing angles required for a speed range of 200 to 1000 rpm.

6. A 2.8 kW, 400 V, 50 Hz, 4-pole, 1370 rpm, delta connected squirrel cage induction motor has the following parameters referred to the stator. $R_s = 2$ Ω, $R'_r = 5$ Ω, $X_s = X'_r = 5$ Ω, $X_m = 80$ Ω. Motor speed is controlled by stator voltage control. When driving a fan load, it runs at rated speed at rated voltage. Calculate:
   (a) Motor terminal voltage, current and torque at 1200 rpm.
   (b) Motor speed, current and torque for the terminal voltage of 300 volts.

7. A three-phase, 420 V, 50 Hz, star connected slip ring induction motor has its speed controlled by means of a GTO chopper in its rotor circuits. The effective phase turn ratio from the rotor to the stator is 0.8. The fitter inductor makes the inductor current to be ripple free losses in rectifier inductor, GTO chopper and the no-load losses of the motor are neglected. The load torque proportional to speed squared is 450 N-m.
   (a) For a minimum motor speed of 1000 rpm, calculate the values of chopper resistance (R).
   (b) For the calculated value of chopper resistance (R), If the speed is raised to 1320 rpm,
      (1) calculate the duty cycle of the chopper
      (2) The inductor current
      (3) Rectified output voltage
      (4) Efficiency in case per phase resistance of stator and rotor are 0.015 Ω and 0.02 Ω respectively.

8. A 440 V, 50 kW, 50 Hz, three-phase slip ring induction motor has the equivalent circuit parameters $r_1 = 0.07\ \Omega$, $r_2 = 0.05\ \Omega$, $X_1 = X_2' = 0.2\ \Omega$, $X_m = 20\ \Omega$. The speed of the motor at rated load is 1420 rpm. Determine the resistance required in the chopper circuit so that the speed can be controlled in the range 1420-1000 rpm at constant torque. Determine the TR for 1100 rpm.

9. A 420 V, 50 Hz, 6-pole star connected to slip ring induction motor speed is controlled by a static Kramer drive. The effective phase ratio from the rotor to the stator is 0.7, and the transformer phase turns ratio from low voltage to high voltage is 0.5. Losses in diode, rectifier, inductor, inverter and transformer were neglected. The load torque proportional to speed squared is 275 Nm at 900 rpm. For a motor operating at 750 rpm, calculate:
   (i) Rotor rectified voltage
   (ii) Inductor current
   (iii) Delay angle of the inverter
   (iv) Efficiency if the inductor resistance is 0.02 Ω and stator and rotor resistance is 0.01 Ω and 0.03 Ω respectively.

10. A three-phase, 4-pole, 50 Hz induction motor has a chopper controlled resistance in the rotor circuit for speed control. The load torque follows a relations $T_L = W^2$ where W is the speed of the motor. When the power switch is on, the torque is 40 Nm at an average slip of 0.04. If $\dfrac{T_{ON}}{T_{OFF}} - 1$, compute the average torque and speed. The motor develops a torque of 75% of on torque when the power switch is off. The speed ranges down to 1400 rpm. Determine the ratio $\dfrac{T_{ON}}{T_{OFF}}$ to give an average torque of 30 Nm.

11. A static Kramer drive is used for speed control of a 4-pole three-phase slip ring induction motor fed from a 415 V, 50 Hz supply. The inverter is connected directly to the supply. If the motor is required to operate at 1200 rpm, find the firing advance angle of the inverter. The voltage across open circuited slip rings at a standstill is 700 V. Allow a voltage drop of 0.7 V and 1.5 V across each of the diodes and thyristors respectively. The voltage drop across the inductor may be neglected.

6

Stator Control of Induction Motor

6.1 INTRODUCTION

The induction motor is the most extensively employed motor in the industry; as it has good self-starting, simple and rugged construction, low cost and reliability, etc. Induction motors are used in many adjustable speed applications which do not require a fast dynamic response. The concept of vector control has make possible that induction motors can be controlled to obtain good dynamic performance as compared to DC or brushless DC motors. The dynamic model of the induction motor is needed for analysis of vector control. It has been investigated that the dynamic model equations developed on a rotating reference frame are simple to describe the characteristics of induction motors. To derive and explain the model of the induction motor is derived and explained in simple terms by using the concept of space vectors and *d-q* variables. There are several techniques to improve performance such as voltage boost, energy savings by voltage reduction, slip compensation, scalar control, vector control, etc. It's nothing to restrict the slip which is built in the motor design. At full speed and full load, an AC motor gets a slip equal to 3% of its synchronous speed based on the motor design and it is not affected by the design of the controller. The slip represents energy loss in the rotor of the AC motor and converts it directly to heat inside the motor which must be dissipated. Because of this fact the AC machine is somewhat limited in terms of full torque thermal speed range.

6.2 SPEED CONTROL OF INDUCTION MOTORS

Induction motors are of two types — squirrel-cage motor and wound-rotor motor (slip ring rotor motor). There are various types of speed control methods of the induction motor. These are:

(1) Pole changing
(2) Stator voltage control
(3) Supply frequency control
(4) Eddy-current coupling
(5) Rotor resistance control
(6) Slip power recovery

Pole changing method is employed for squirrel-cage motor whereas stator voltage control, supply frequency control and eddy-current coupling methods are applicable for both wound-rotor induction motor and squirrel-cage induction motor. Rotor resistance control and slip power recovery methods are used for speed control of wound-rotor induction motor.

6.2.1   Pole Changing

For a given frequency speed is inversely proportional to the number of poles. Synchronous speed, and therefore, motor speed can be varied by varying the number of poles. Arrangement for changing of number of poles has to be included at the manufacturing stage, and such a machine is called pole changing motor or multi-speed motor.

In squirrel-cage rotor induction motor, the number of poles is same as that of the stator winding. So there is no requirement for changing the number of poles. But for wound-rotor induction motor provision for changing the number of poles in the rotor is needed, which complicates the machine. So it is only employed for squirrel-cage induction motor.

A simple but costly provision for changing the number of stator poles is to use two separate windings which are wound for two different pole numbers. An economical and common substitute is to use single stator winding divided into a few coil groups. By dividing the winding into a number of coil groups and bringing out terminals of these groups, the number of arrangements of different pole numbers can be obtained.

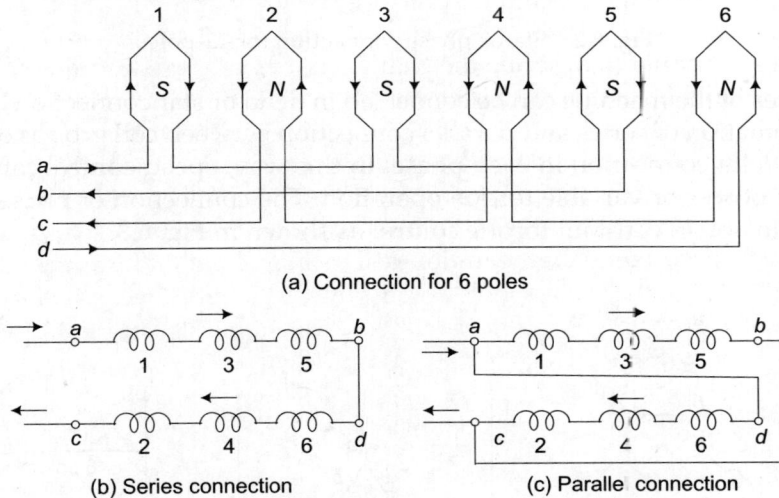

(a) Connection for 6 poles

(b) Series connection   (c) Paralle connection

Fig. 6.1   Stator phase connection for 6 poles.

Figure 6.1 shows a phase winding consisting of six coils divided into two groups given in (a-b) and (c-d). Stator phase connection having odd numbered coils (1, 3, 5) connected in series in a-b and even numbered coils (2, 4, 6) connected in series in c-d given in Fig. 6.1(b). The coils arranged are to carry currents in the given directions by connecting coil groups either in series or parallel as shown in Fig. 6.1(b) and 6.1(c). The machine has six poles. If the current through the coil group a-b is reversed as shown in Fig. 6.2, then all coils will produce north poles. Fluxes coming out of the north poles will now find paths through interpol of spaces for going out as a result it produces south poles in interpole spaces. The machine will now have 12 poles, and again the direction of current through coils can be obtained by connecting two sections a-b and c-d either in series or in parallel for both pole numbers 6 and 12.

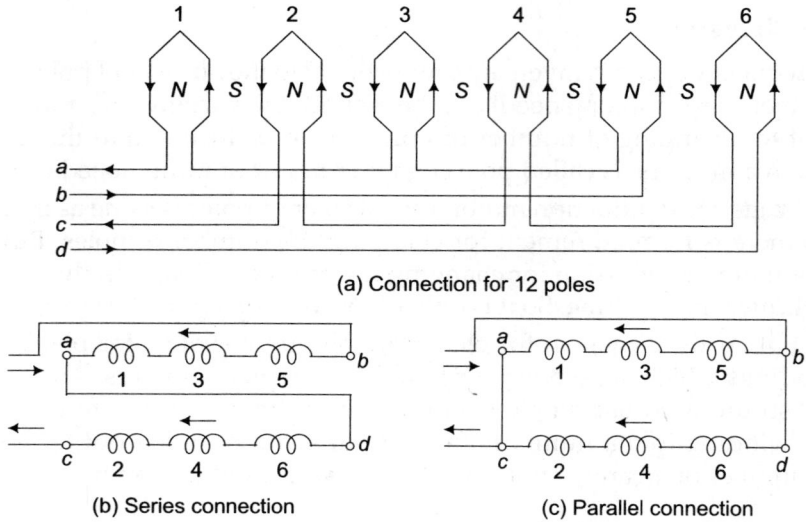

(a) Connection for 12 poles

(b) Series connection

(c) Parallel connection

Fig. 6.2   Stator phase connection for 12-poles.

Three phases of the machine can be connected in delta or star connection by choosing a suitable combination of series and parallel connection between coil groups of each phase, and star and delta connection in each phase. In this way, speed control can be obtained with constant power or variable torque operation. The connection of phases and speed-torque curve to obtain constant torque control as shown in **Fig. 6.3**.

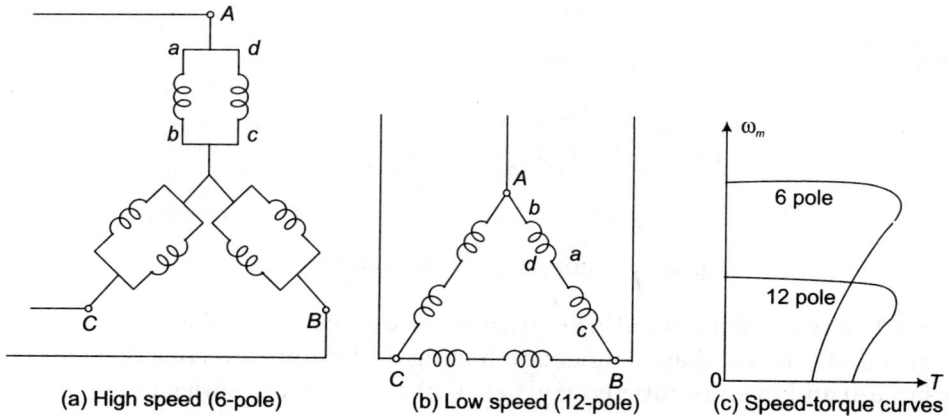

(a) High speed (6-pole)

(b) Low speed (12-pole)

(c) Speed-torque curves

Fig. 6.3   Constant torque control.

The connection of phases and speed-torque curves to obtain constant power control as shown in **Fig. 6.4**.

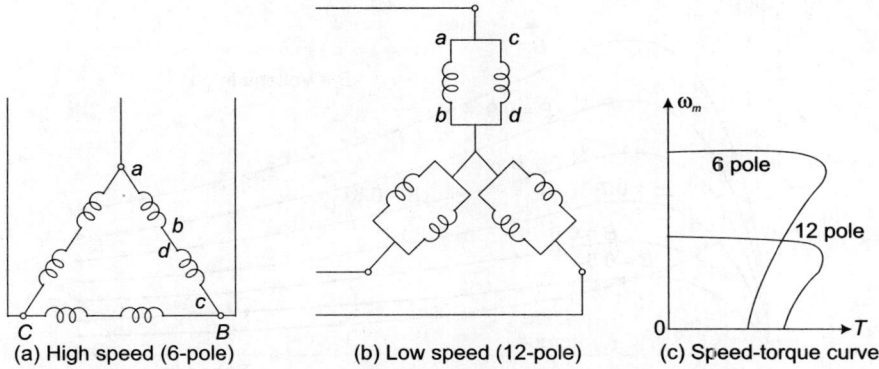
Fig. 6.4   Constant power control.

Connections and speed-torque curves for operations of variable torque control is shown in **Fig. 6.5.**

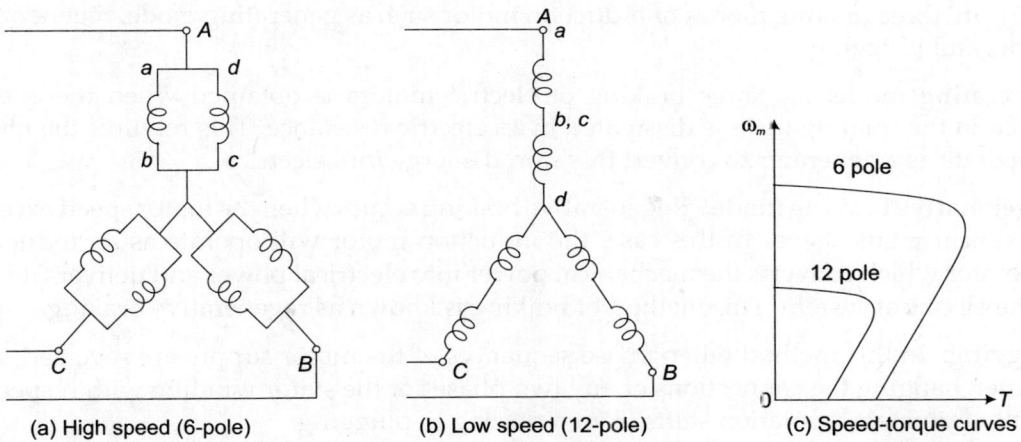

(a) High speed (6-pole)   (b) Low speed (12-pole)   (c) Speed-torque curves

Fig. 6.5   Variable torque control.

6.3   STATOR VOLTAGE CONTROL

It is a slip control method with constant frequency variable voltage being supplied to the motor stator windings. Of course, the voltage must only be reduced below the rated value. For a motor operating at full load slip, if the slip is to be doubled for constant load torque then the voltage must be decreased by a factor of $\dfrac{1}{\sqrt{2}}$ and the corresponding current increases to $\sqrt{2}$ of the full load value. Therefore, motor gets overheated. Thus, this method is not applicable for speed control. It has a limited application for motor driving fan type load whose torque requirement is proportional to the square of speed. It is a commonly used method for ceiling fans driven by single-phase induction motors that have large standstill impedance restricting the current drawn by the stator.

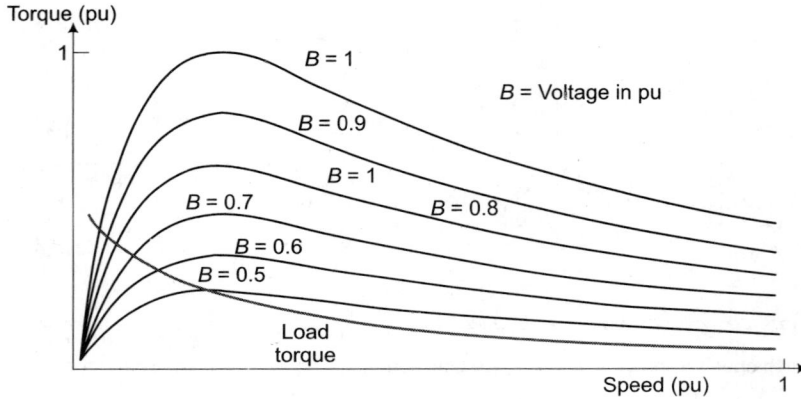

Fig. 6.6   Torque-speed characteristics of stator voltage control.

Stator voltage control means that the speed of induction motor can be changed by changing the stator voltage as the torque is proportional to the square of the voltage. There are three braking modes of induction motor such as generating mode, regenerative mode, and plugging.

Generating mode: Dynamic braking of electric motors is obtained when the energy stored in the rotating mass is dissipated in an electric resistance. This requires the motor to operate as a generator to convert this stored energy into electrical.

Regenerative braking mode: Regenerative braking occurs when the motor speed exceeds the synchronous speed. In this case, the induction motor will operate as an induction generator which converts the mechanical power into electrical power and delivers it back to the electrical system. This method of braking is known as regenerative braking.

Plugging: In this method when phase sequences of the motor supply are reversed, that is, interchanging the connections of any two phases of the stator winding with respect to supply terminals, operation shifts from motoring to plugging.

$$T_{em} = \frac{3R'_r}{s\omega_s} \frac{V_s^2}{\left(R_s + \dfrac{R'_r}{s}\right)^2 + (X_{ls} + X'_{lr})^2}$$

$$T_{em} = \frac{V_s}{\left(R_s + \dfrac{R'_r}{s}\right)^2 + (X_{ls} + X'_{lr})}$$

In the regenerating region, the machine acts as a generator. The rotor moves at a super synchronous speed in the same direction as that of the air gap flux so that the slip goes to negative, creating negative or regenerating torque. With a variable-frequency power supply, the machine stator frequency can be controlled to be lower than the rotor speed ($\omega_e < \omega_r$,) to get a regenerative braking effect.

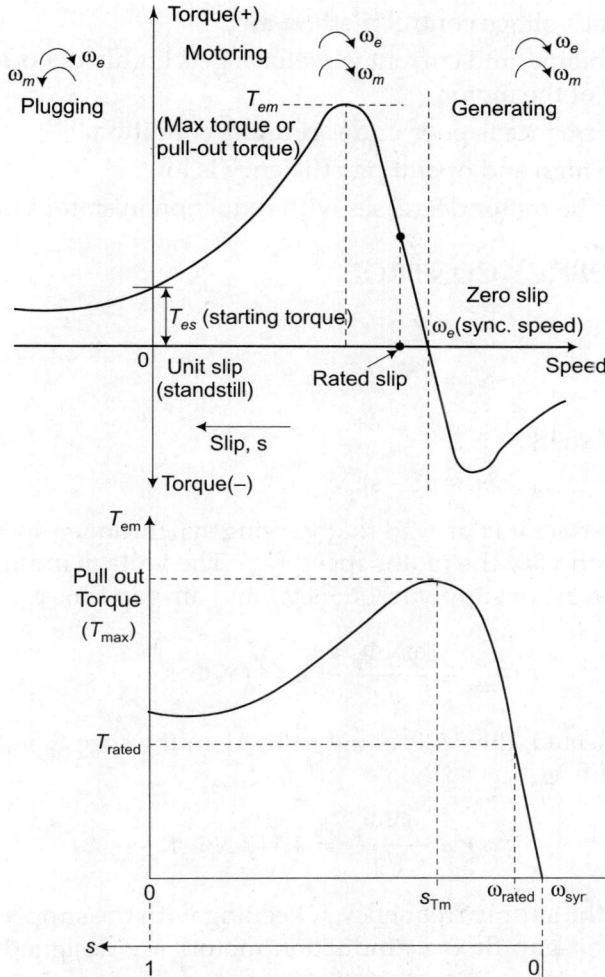

Fig. 6.7   Torque-speed characteristics torque-slip characteristics of induction motor.

The application of the stator voltage control method is applicable for torque demand reduced with speed.

Examples:

- Fan
- Pump drives

  The merits of stator voltage control method are:

- The simple control circuitry
- Compact size
- Fast response time

The demerits of stator voltage control method are:
- Due to harmonics, voltage, and current waveform gets highly distorted, and therefore, affects the efficiency of the motor.
- At low speeds, performance is poor under running condition.
- Resistance losses are high and operating efficiency is low.
- Maximum torque of the motor decreases with reduction in stator voltage.

6.4   SUPPLY FREQUENCY CONTROL

Synchronous speed

$$N_s = 120\,\frac{f}{p}.$$

and, motor speed is given by:

$$N_r = (1 - s)\,N_s$$

From the above equation, it is proved that varying synchronous speed by varying the supply frequency (f) can vary the motor speed (N_r). The voltage induced in the stator is proportional to the product of supply frequency f_s and air-gap flux φ_m.

$$E_{rms} = \frac{\omega N\Phi_p}{\sqrt{2}} = 4.44\,f\,N\Phi_p K_w$$

If stator drop is neglected, then E is equal to V. Then the supply voltage will become proportional to f_s and flux φ_m.

$$E_{rms} = \frac{\omega N\Phi_p}{\sqrt{2}} = 4.44\,f\,N\Phi_p K_w$$

Any degradation in the supply frequency f_s, keeping with the supply voltage constant, causes the increase of air-gap flux φ_m. Induction motors are designed to operate at the knee point of the magnetization characteristic to make a full use of the magnetic material. Therefore, the increase in flux will saturate the motor. This will increase the magnetizing current and distort the line current and voltage, increase in core loss and stator I^2R loss and produce a high-pitch acoustic noise. Also, a decrease in flux is undesirable avoided to retain the torque capability of the motor. Therefore, variable frequency control below the rated frequency is generally carried out at rated air-gap flux by varying supply voltage with frequency so as to keep $\dfrac{V}{f}$ ratio constant at the rated value.

It appears that the rotor will rotate at a speed slightly lower than the stator frequency (slip). Hence, a speed control is obtained when the stator frequency is varied. For maintaining constant flux, the V/F ratio should be kept constant.

$$T_{em} = \frac{3R'_r}{s\omega_s}\,\frac{V_s^2}{\left(R_s + \dfrac{R'_r}{s}\right)^2 + (X_{ls} + X'_{lr})^2}$$

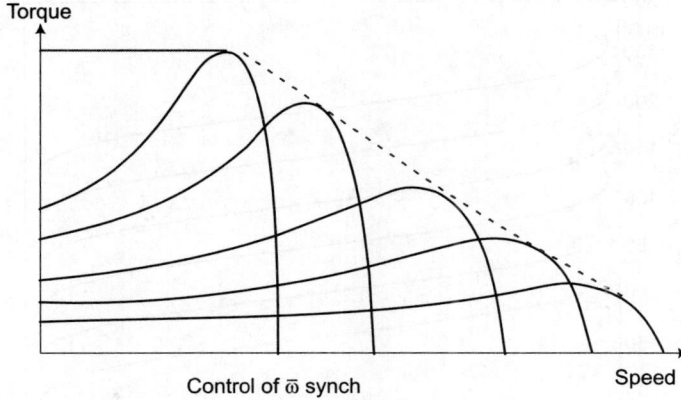

Fig. 6.8   Torque-speed characteristics of constant torque.

6.5   SCALAR AND VECTOR CONTROL OF INDUCTION MOTOR

Scalar control of an AC motor drive is possible due to variation in the magnitude of the control variables. As the name implies the vector control involves the variation of both the magnitude and phase of the control variables.

Voltage can be used to control the air-gap flux and frequency, or slip can be used to control the torque. However, flux and torque are the functions of frequency and voltage respectively, but this case is ignored in scalar control.

Scalar control gives inferior dynamic performance of an AC motor as compared to vector control, but its implementation is simpler. In variable-speed applications with a small variation in motor speed and tolerable loading, a scalar control system can provide satisfactory performance. However, if precision control is required, then a vector control system must be employed.

6.5.1   Constant Air-Gap Flux

Generally, an induction motor requires nearly constant amplitude of air-gap flux for adequate performance of the motor. Since the air-gap flux is the integral of the voltage applied across the magnetizing inductance, and assuming that the air-gap voltage is sinusoidal.

$$\lambda_{ag} = \int v_{ag}\, dt = \int V_{ag} \sin \omega t\, dt = -\frac{V_{ag}}{\omega} \cos \omega t$$

Thus, a constant v/f ratio results in a constant air-gap flux. The torque-speed curves with a constant air-gap flux at different excitation frequencies are shown in **Fig. 6.9(a)**.

From the curves, it can be seen that we will obtain the same torque at the same value of slip speed if motor operates at a constant air-gap flux. This type of control may be implemented either in open loop or in a closed loop. A set of 6-step voltage waveforms illustrating constant v/f is shown in **Fig. 6.9(b)**.

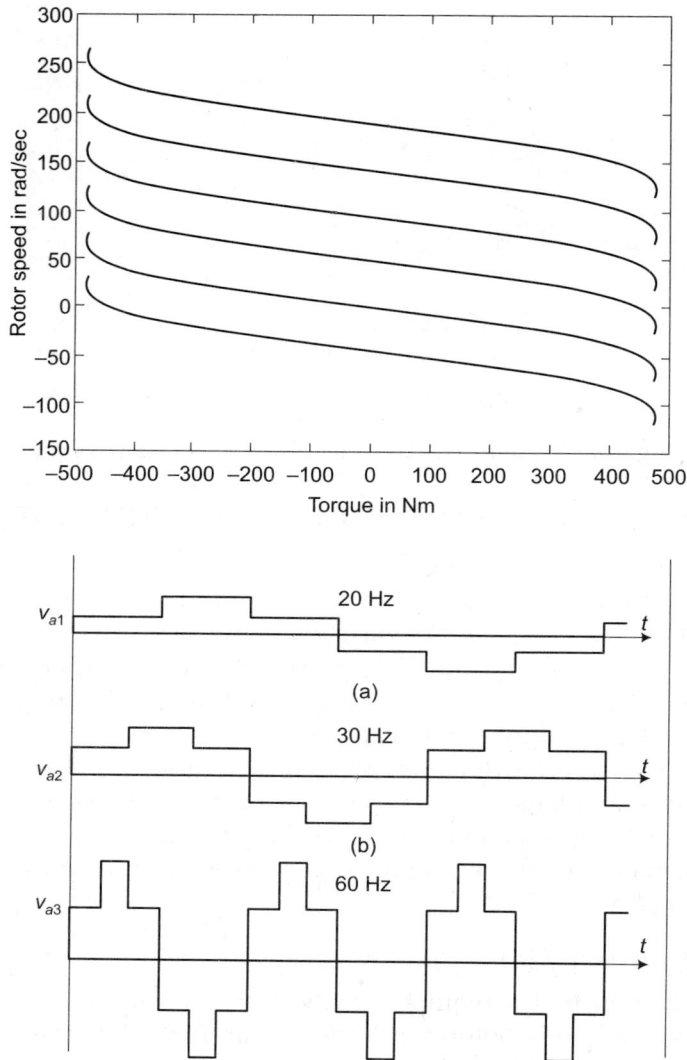

Fig. 6.9   (a) & (b) In constant air-gap flux torque-speed curve pulse with respect to voltage-time.

6.6   AC VOLTAGE CONTROLLERS

AC voltage controllers (AC line voltage controllers) are used to change the rms value of the alternating voltage impressed to a load circuit by providing thyristors between the load and a constant voltage AC source. The rms value of alternating voltage is controlled by adjusting the triggering angle (α) of the thyristors in the AC voltage controller circuits.

In other words, an AC voltage controller is a thyristor power converter which is used to convert a fixed voltage, fixed frequency AC input supply to obtain a variable voltage AC output.

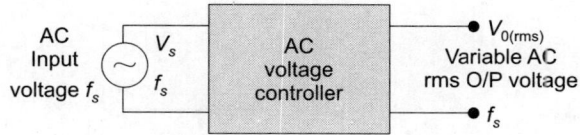

Fig. 6.10   Two-port N/w of AC voltage controller.

There are two different types of thyristor controls used to control the AC power flow.

- On-off control
- Phase control

In on-off control technique thyristors act as switches to connect the load circuit to the input AC supply (source) for a few cycles and then to disconnect it for a few input cycles. The thyristors thus act as a high-speed contactor (or high-speed AC switch).

6.6.1   Type of AC Voltage Controllers

Based on the type of input AC supply given to the circuit, the AC voltage controllers are classified into two types:

- Single-phase AC voltage controllers.
- Three-phase AC voltage controllers.

In India single-phase AC voltage controllers operate with a single-phase AC supply voltage of 230 V rms at 50 Hz. Three-phase AC voltage controllers operate with three-phase AC supply of 440 V rms at 50 Hz supply frequency.

Each type of controller may be subdivided as:

- Unidirectional or half-wave AC voltage controller.
- Bidirectional or full-wave AC voltage controller.

In brief, different types of AC voltage controllers are:

- Single-phase half-wave AC voltage controller (unidirectional controller).
- Single-phase full-wave AC voltage controller (bidirectional controller).
- Three-phase half-wave AC voltage controller (unidirectional controller).
- Three-phase full-wave AC voltage controller (bidirectional controller).

6.6.2   Principle of on-off Control Technique (Integral Cycle Control)

The basic principle of on-off control technique is explained with reference to a single-phase full-wave AC voltage controller circuit shown in **Fig. 6.11**. The thyristor switches, T_1 and T_2, are turned on by applying appropriate gate trigger pulses to connect the input AC supply to the load for 'n' number of input cycles during the time interval t_{ON} The thyristor switches, T_1 and T_2, are turned off by blocking the gate trigger pulses for 'm' number of input cycles during the time interval t_{OFF}. The AC controller ON time t_{ON} usually consists of an integral number of input cycles.

$$R = R_L = \text{Load resistance}$$

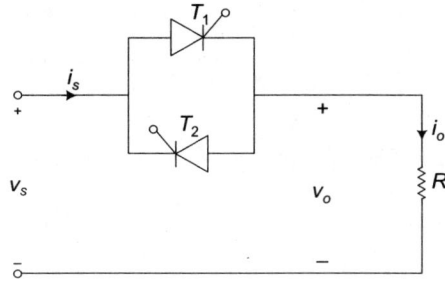

Fig. 6.11   Single-phase full-wave AC voltage controller circuit.

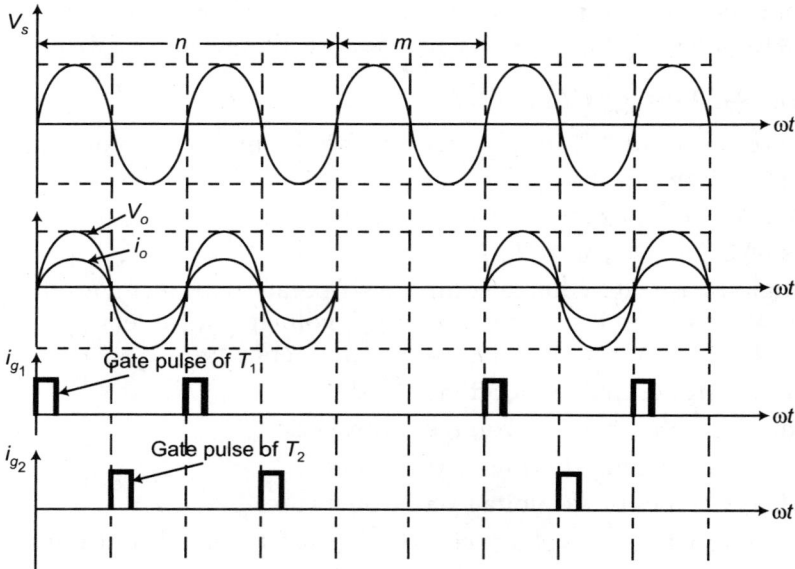

Fig. 6.12   Single-phase full-wave AC voltage controller waveform.

n = Two input cycles. Thyristors are turned ON during t_{ON} for two input cycles.

m = One input cycle. Thyristors are turned OFF during t_{OFF} for one input cycle

Thyristors are turned on precisely at the zero voltage crossings of the input supply. The thyristor T_1 is turned on at the beginning of each positive half cycle by applying the gate trigger pulses to T_1 as shown, during the on time t_{ON}. The load current flows in the positive direction, which is the downward direction as shown in the circuit diagram when T_1 conducts. The thyristor T_2 is turned on at the beginning of each negative half cycle, by applying a gating signal to the gate of T_2, during t_{ON}. The load current flows in the reverse direction, which is the upward direction when T_2 conducts. Thus, we obtain a bidirectional load current flow (alternating load current flow) in a AC voltage controller circuit, by triggering the thyristors alternately.

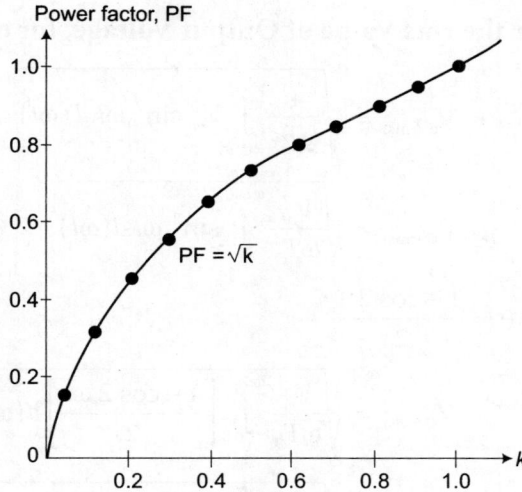

Fig. 6.13   Power actor curve.

This type of control is used in applications which have high mechanical inertia and high thermal time constant (industrial heating and speed control of AC motors). Due to zero voltage and zero current switching of thyristors, the harmonics generated by switching actions are reduced.

For a sine wave input supply voltage,

$$v_s = V_m \sin \omega t = \sqrt{2}\, V_S \sin \omega t$$

V_S = rms value of input AC supply = $\dfrac{V_m}{\sqrt{2}}$ = rms phase supply voltage.

If the input AC supply is connected to load for 'n' number of input cycles and disconnected for 'm' number of input cycles, then

$$t_{ON} = n \times T, \quad t_{OFF} = m \times T$$

where
$$T = \frac{1}{f} \text{ input cycle time (time period)}$$

f = input supply frequency.

t_{ON} = controller on time = $n \times T$.

t_{OFF} = controller off time = $m \times T$

T_O = Output time period = $(t_{ON} + t_{OFF}) = (nT + mT)$.

We can show that,

Output rms voltage $V_O(\text{rms}) = V_{i(\text{rms})} \sqrt{\dfrac{t_{ON}}{T_O}} = V_s \sqrt{\dfrac{t_{ON}}{T_O}}$

where $V_i(\text{rms})$ the rms is input supply voltage = V_S.

6.6.3   Expression for the rms Value of Output Voltage, for on-off Control Method

Output rms voltage $V_{O\,(rms)} = \sqrt{\dfrac{1}{\omega T_O} \displaystyle\int_{\omega t=0}^{\omega t_{ON}} V_m^2 \sin^2 \omega t.d\,(\omega t)}$

$$V_{O\,(rms)} = \sqrt{\dfrac{V_m^2}{\omega T_O} \int_{0}^{\omega t_{ON}} \sin^2 \omega t.d\,(\omega t)}$$

Substituting for $\sin^2 \theta = \dfrac{1 - \cos 2\theta}{2}$

$$V_{O\,(rms)} = \sqrt{\dfrac{V_m^2}{\omega T_O} \int_{0}^{\omega t_{ON}} \left[\dfrac{1 - \cos 2\omega t}{2}\right] d\,(\omega t)}$$

$$V_{O\,(rms)} = \sqrt{\dfrac{V_m^2}{2\omega T_O} \left[\int_{0}^{\omega t_{ON}} d\,(\omega t) - \int_{0}^{\omega t_{ON}} \cos 2\omega t.d\,(\omega t)\right]}$$

$$V_{O\,(rms)} = \sqrt{\dfrac{V_m^2}{2\omega T_O} \left[(\omega t)\Big/_0^{\omega t_{ON}} - \dfrac{\sin 2\omega t}{2}\Big/_0^{\omega t_{ON}}\right]}$$

$$V_{O\,(rms)} = \sqrt{\dfrac{V_m^2}{2\omega T_O} \left[(\omega t_{ON} - 0) - \dfrac{\sin 2\,\omega t_{ON} - \sin 0}{2}\right]}$$

Now t_{ON} = An integral number of input cycles. Hence,

$t_{ON} = T, 2T, 3T, 4T, 5T, \ldots,$ & $\omega t_{ON} = 2\pi, 4\pi, 6\pi, 8\pi, 10\pi, \ldots.$

where T is the input supply time period (T = input cycle time period). Thus, we note that $\sin 2\omega t_{ON} = 0$

$$V_{O\,(rms)} = \sqrt{\dfrac{V_m^2 \,\cancel{\omega}\, t_{ON}}{2\,\cancel{\omega}\, T_O}} = \dfrac{V_m}{\sqrt{2}} \sqrt{\dfrac{t_{ON}}{T_O}}$$

$$V_{O\,(rms)} = V_{i(rms)} \sqrt{\dfrac{t_{ON}}{T_O}} = V_S \sqrt{\dfrac{t_{ON}}{T_O}}$$

where $V_i(rms) = \dfrac{V_m}{\sqrt{2}} = V_S$ = rms value of input supply voltage;

$$\dfrac{t_{ON}}{T_O} = \dfrac{t_{ON}}{t_{ON} + t_{OFF}} = \dfrac{nT}{nT + mT} = \dfrac{n}{(n+m)} = k = \text{duty cycle } (d).$$

$$V_{O\,(rms)} = V_S \sqrt{\dfrac{n}{(m+n)}} = V_S \sqrt{k}$$

6.6.4   Parameters of AC Voltage Controller

- **rms output (load) voltage**

$$V_{O\,(rms)} = \left[\frac{n}{2\pi\,(n+m)} \int_0^{2\pi} V_m^2 \sin^2 \omega t.d\,(\omega t) \right]^{1/2}$$

$$V_{O\,(rms)} = \frac{V_m}{\sqrt{2}} \sqrt{\frac{n}{(m+n)}} = V_{i(rms)}\sqrt{k} = V_s\sqrt{k}$$

$$V_{O\,(rms)} = V_{i(rms)}\sqrt{k} = V_s\sqrt{k}$$

where $V_S = V_{i\,(rms)}$ = rms value of input supply voltage.

- **Duty cycle**

$$k = \frac{t_{ON}}{T_O} = \frac{t_{ON}}{(t_{ON}+t_{OFF})} = \frac{nT}{(m+n)T}$$

where, $k = \dfrac{n}{(m+n)}$ duty cycle (d).

- **rms load current**

$$I_{O\,(rms)} = \frac{V_{O(rms)}}{Z} = \frac{V_{O(rms)}}{R_L}\; ;\text{ for a resistive load } Z = R_L.$$

- **Output AC (load) power**

$$P_O = I_O^2\,(rms) \times R_L$$

- **Input power factor**

$$PF = \frac{P_O}{VA} = \frac{\text{output load power}}{\text{input supply volt amperes}} = \frac{P_O}{V_S I_S}$$

$$PF = \frac{I_{O(rms)}^2 \times R_L}{V_{i(rms)} \times I_{in(rms)}}\; ; I_S = I_{in\,(rms)} = \text{rms input supply current.}$$

The input supply current is same as the load current $I_{in} = I_O = I_L$

Hence, rms supply current = rms load current; $I_{in\,(rms)} = I_{O\,(rms)}$.

$$PF = \frac{I_{O(rms)}^2 \times R_L}{V_{i(rms)} \times I_{in\,(rms)}} = \frac{V_{O\,(rms)}}{V_{i(rms)}} = \frac{V_{i(rms)}\sqrt{k}}{V_{i(rms)}} = \sqrt{k}$$

$$PF = \sqrt{k} = \sqrt{\frac{n}{m+n}}$$

- **The average current of thyristor $I_{T(Avg)}$**

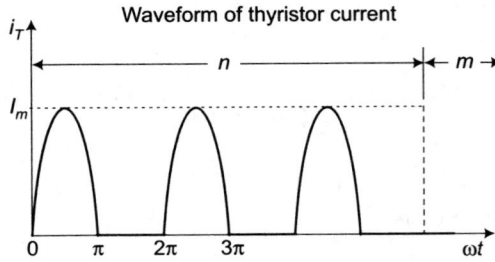

Waveform of thyristor current

$$I_{T(Avg)} = \frac{n}{2\pi(m+n)} \int_0^\pi I_m \sin \omega t . d(\omega t)$$

$$I_{T(Avg)} = \frac{nI_m}{2\pi(m+n)} \int_0^\pi \sin \omega t . d(\omega t)$$

$$I_{T(Avg)} = \frac{nI_m}{2\pi(m+n)} \left[-\cos \omega t \Big/ \begin{array}{c} \pi \\ 0 \end{array} \right]$$

$$I_{T(Avg)} = \frac{nI_m}{2\pi(m+n)} \left[-\cos \pi + \cos 0 \right]$$

$$I_{T(Avg)} = \frac{nI_m}{2\pi(m+n)} \left[-(-1) + 1 \right]$$

$$I_{T(Avg)} = \frac{n}{2\pi(m+n)} \left[2I_m \right]$$

$$I_{T(Avg)} = \frac{I_m n}{\pi(m+n)} = \frac{k . I_m}{\pi}$$

$$k = \text{duty cycle} = \frac{t_{ON}}{(t_{ON} + t_{OFF})} = \frac{n}{(n+m)}$$

$$I_{T(Avg)} = \frac{I_m n}{\pi(m+n)} = \frac{k . I_m}{\pi},$$

where $I_m = \dfrac{V_m}{R_L}$ = maximum or peak thyristor current.

- **rms current of thyristor $I_{T(rms)}$**

$$I_{T(rms)} = \left[\frac{n}{2\pi(n+m)} \int_0^\pi I_m^2 \sin^2 \omega t . d(\omega t) \right]^{1/2}$$

$$I_{T(rms)} = \left[\frac{nI_m^2}{2\pi(n+m)} \int_0^\pi \sin^2 \omega t . d(\omega t) \right]^{1/2}$$

$$I_{T(rms)} = \left[\frac{nI_m^2}{2\pi(n+m)} \int_0^\pi \frac{(1 - \cos 2\omega t)}{2} d(\omega t) \right]^{1/2}$$

Q.2 A three-phase, 50 hp, 460 V, 60 Hz, 865 rpm induction motor is operating at rated conditions and has $P_T = 43.75$ kW and $I_1 = 61$ A. It is known that $R_1 = 0.15\,\Omega$ and rotational losses at rated speed are 1050 W. Determine (a) full-load power factor, (b) full-load efficiency, (c) total rotor coil ohmic losses, and (d) total core losses.

Solution:

(a) $PF = \dfrac{P_T}{\sqrt{3}\,V_L I_1} = \dfrac{43,750}{\sqrt{3}\,(460)(61)} = 0.9$ lagging

(b) $\eta = \dfrac{100\,P_s}{P_T} = \dfrac{100\,(50\times746)}{43,750} = 85.26\%$

(c) $s = \dfrac{n_s - n_m}{n_s} = \dfrac{900 - 865}{900} = 0.03889$

$3P_g = \dfrac{3P_d}{1-s} = \dfrac{P_s + P_{FW}}{1-s} = \dfrac{50\,(746) + 1050}{1 - 0.03889} = 39,901.8$ W

$3\,(I_2')^2\,R_2' = s\,(3\,P_g) = 0.03889\,(39,901.8) = 1551.8$ W

(d) $3P_c = P_T - 3I_1^2\,R_1 - 3P_g$

$3P_c = 43,750 - 3(61)^2\,(0.15) - 39,901.8 = 2173.7$ W

Q.3 A three-phase, 6-pole, 10 hp, 400 Hz induction motor has a slip of 3% at rated output power. Friction and windage losses are 300 W at rated speed. The rated condition total core losses are 350 W. $R_1 = R_2' = 0.05\,\Omega$. $X_1 = X_2' = 0.15\,\Omega$. If the motor is operating at rated output power, speed, and frequency, find (a) rotor speed, (b) frequency of rotor currents, (c) total power across the air gap, (d) efficiency, and (e) applied line voltage. Use the approximate equivalent **circuit for analysis.**

Solution:

(a) $n_s = \dfrac{120f}{p} = \dfrac{(120)(400)}{6} = 8000$ rpm

$n_m = (1 - s)\,n_s = (1 - 0.03)\,(8000) = 7760$ rpm

(b) $f_r = s_f = (0.03)\,(400) = 12$ Hz

(c) $3P_d = P_s + P_{FW} = (10)\,(746) + 300 = 7760$ W

$3P_g = \dfrac{3P_d}{(1-s)} = \dfrac{7760}{1 - 0.03} = 8000$ W

(d) The reflected secondary current is found by

$I_2' = \left[\dfrac{sP_g}{R_2'}\right]^{1/2} = \sqrt{\dfrac{(0.03)(8000/3)}{0.05}} = 40$ A

$\text{Losses} = 3\,(I_2')^2\,(R_1 + R_2') + 3P_c + P_{FW}$

$= 3\,(40)^2\,(0.05 + 0.05) + 350 + 300 = 1130$ W

$$\eta = \frac{P_s(100)}{P_s + \text{losses}} = \frac{(10)(746)(100)}{(10)(746) + 1130} = 88.94\%$$

(e)
$$V_1 = I_2' \left| R_1 + \frac{R_2'}{s} + jX_{eq} \right| = 40 \left| 0.05 + \frac{0.05}{0.03} + j0.3 \right| = 69.71 \text{ V}$$

$$V_L = \sqrt{3} V_1 = \sqrt{3}(69.71) = 120.7 \text{ V}$$

Q.4 A 4-pole, 230 V, three-phase induction motor has a value of secondary resistance such that the motor produces maximum developed torque at stall. Neglect core losses and use the Thevenin equivalent circuit for analysis. (See **Fig. 6.56**). Known equivalent circuit values are:

$$R_1 = 0.2 \qquad\qquad R_2 = 1.1064$$
$$X_1 = 0.5 \ \Omega \qquad\qquad X_m = 20 \ \Omega$$

Find (a) the reflected value of X_2 and (b) the total developed torque at stall.

Solution:

(a) Based on [6.50] with, $R_c \to \infty$,

$$Z_{Th} = \frac{-X_m X_1 + jX_m R_1}{R_1 + j(X_1 + X_m)} = \frac{-(20)(0.5) + j(20)(0.2)}{0.2 + j(0.5 + 20)}$$

$$Z_{Th} = R_{Th} + jX_{Th} = 0.19 + j0.49 \ \Omega$$

Since $s_{max+} = 1$, [6.56] leads to

$$\sqrt{R_{Th}^2 + (X_{Th} + X_2')^2} = R_2' \text{ hence,}$$

$$(X_2') + 2 X_{Th} X_2' + [X_{Th}^2 + R_{Th}^2 - (R_2')^2] = 0$$

Substitute known values and apply the quadratic formula.

$$(X_2')^2 + 0.98 X_2' - 0.9479 = 0$$

$$X_2' = 0.6 \ \Omega$$

where the negative value was discarded as extraneous.

(b) By [6.49] for, $R_c \to \infty$,

$$V_{Th} = \frac{X_m V_1}{\sqrt{R_1^2 + (X_1 + X_m)^2}} = \frac{(20)(230/\sqrt{3})}{\sqrt{(0.2)^2 + (0.5 + 20)^2}} = 129.55 \text{ V}$$

$$\omega_s = \frac{2}{p} \omega = \frac{2}{4}(2\pi \times 60) = 188.49 \text{ rad/s}$$

$$3 T_{d\,max+} = \frac{3 V_{Th}^2}{2\omega_s \left[R_{Th} + \sqrt{R_{Th}^2 + (X_{Th} + X_2')^2} \right]} = \frac{3(129.55)^2}{2(188.49)\left[0.19 + \sqrt{(0.19)^2 + (1.09)^2} \right]}$$

$$3 T_{d\,max+} = 103.02 \text{ N.m}$$

Q.5 A three-phase, 50 hp, 460 V, 60 Hz, 865 rpm induction motor is operating at rated conditions and has $P_T = 43.75$ kW and $I_1 = 61$ A. It is known that $R_1 = 0.15\ \Omega$ and rotational losses at rated speed are 1050 W. Determine (a) full-load power factor, (b) full-load efficiency, (c) total rotor coil ohmic losses, and (d) total core losses.

Solution:

(a)
$$PF = \frac{P_T}{\sqrt{3}\,V_L\,I_1} = \frac{43{,}750}{\sqrt{3}\,(460)\,(61)} = 0.9 \text{ lagging}$$

(b)
$$\eta = \frac{100\,P_s}{P_T} = \frac{100\,(50 \times 746)}{43{,}750} = 85.26\%$$

(c)
$$s = \frac{n_s - n_m}{n_s} = \frac{900 - 865}{900} = 0.03889$$

$$3\,P_g = \frac{3P_d}{1-s} = \frac{P_s + P_{FW}}{1-s} = \frac{50\,(746) + 1050}{1 - 0.03889} = 39{,}901.8 \text{ W}$$

$$3\,(I_2')^2\,R_2' = s\,(3\,P_g) = 0.03889\,(39{,}901.8) = 1551.8 \text{ W}$$

(d)
$$3\,P_c = P_T - 3I_1^2\,R_1 - 3\,P_g$$
$$3\,P_c = 43{,}750 - 3(61)^2\,(0.15) - 39{,}901.8 = 2173.7 \text{ W}$$

Q.6 A three-phase, 4-pole, 600 V, 60 Hz induction motor is modelled by $Z_{Th} = 0.6933 + j1.933\ \Omega$, $R_2' = 4.5\ \Omega$, and $X_2' = 2\ \Omega$. Find the shaft speed at which maximum torque occurs if the motor is absorbs power from the three-phase lines at rated frequency.

Solution:

$$s_{max} = \frac{R_2'}{\sqrt{R_{Th}^2 + (X_{Th} + X_2')^2}} = \frac{4.5}{\sqrt{(0.6933)^2 + (3.933)^2}} = 1.127$$

$$n_s = \frac{120\,f}{p} = \frac{120\,(60)}{4} = 1800 \text{ rpm}$$

$$n_m = (1 - s)\,n_s = (1 - 1.127)\,1800 = -2282 \text{ rpm}$$

This motor is operating in the braking or plugging mode.

Q.7 A three-phase, 230 V induction motor is operating at the no-load condition with rated voltage applied. Equivalent circuit parameters are:

$$R_1 = 0.26\ \Omega \qquad R_2' = 0.4\ \Omega \qquad R_c = 143\ \Omega$$
$$X_1 = 0.6\ \Omega \qquad X_2' = 20\ \Omega \qquad X_m = 22.2\ \Omega$$

It is known that the rated voltage core losses are equal to the rotational losses. Assume that for this no-load condition the coil resistive voltage drops and the leakage reactance voltage drops can be neglected. Determine (a) no-load slip, (b) no-load input power factor, and (c) no-load line current.

Solution:

(a) Under the stated assumption, the per phase equivalent circuit can be drawn as shown by **Fig. 6.23**. Since $3P_c = P_{FW}$, and reasoning that the converted power must be the rotational losses,

$$\frac{V_1^2}{R_c} = \frac{V_1^2}{R_2'(1-s)/s}$$

When,

$$s = \frac{R_2'}{R_2' + R_c} = \frac{0.4}{0.4 + 143} = 0.00279$$

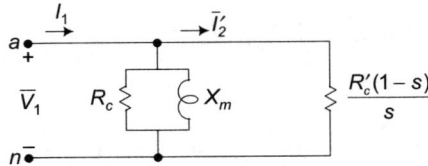

Fig. 6.23

(b) From (1), it is seen that $R_c = R_2'(1-s)/s$. Thus,

$$Z_{in} = \frac{\left(\dfrac{R_c}{2}\right)(jX_m)}{\dfrac{R_c}{2} + jX_m} = \frac{\left(\dfrac{143}{2}\right)(22.2\angle 90°)}{\dfrac{143}{2} + j22.2} = 21.2\angle 72.75° \ \Omega$$

$$PF_{NL} = \cos\left(\angle Z_{in}\right) = \cos\left(72.75°\right) = 0.297 \text{ lagging}$$

(c)
$$I_1 = \frac{V_1}{Z_{in}} = \frac{230/\sqrt{3}}{21.2} = 6.26 \text{ A}$$

Multiple-Choice Questions

1. In stator voltage control method of the induction motor. The torque is
   (a) Directly proportional to voltage
   (b) Directly proportional to voltage square
   (c) Inversely proportional to voltage
   (d) Inversely proportional to voltage square

2. Mathematical expression for rotor speed in rpm is:
   (a) $n_m = n_s(1-s)$ (b) $n_m = n_s(s-1)$

   (c) $n_m = n_s \dfrac{1}{s}$ (d) $n_m = n_s \dfrac{(1-s)}{s}$

3. In the stator voltage control method the relationship between resistance losses and operating efficiency.
   (a) Both are high
   (b) Both are low
   (c) High and low
   (d) Low and high

4. In which methods, the induction motor acts as a DC series motor.
   (a) Stator voltage control
   (b) Variable frequency control
   (c) Eddy current coupling
   (d) Rotor resistance control

5. In V/f control of induction motor the voltage-frequency curve is
   (a) Step function
   (b) Parabolic function
   (c) Ramp function
   (d) Impulse function

6. Pole changing method of speed control is used in
   (a) Slip ring induction motor
   (b) DC shunt motor
   (c) DC series motor
   (d) Squirrel-cage induction motor

7. Which of the following pair is used for frequency converters?
   (a) Wound-rotor induction motor and synchronous motor
   (b) Wound-rotor induction motor and squirrel-cage induction motor
   (c) Squirrel-cage induction motor and synchronous motor
   (d) Any of the above

8. Most commonly used AC motor is:
   (a) Synchronous motor
   (b) Slip ring induction motor
   (c) Squirrel-cage induction motor
   (d) AC commutator induction motor

9. The motor commonly used in computers digital systems is
   (a) DC shunt motor
   (b) Induction motor
   (c) Stepper motor
   (d) Synchronous motor

10. A reluctance motor on overload runs as
    (a) Synchronous motor
    (b) DC motor
    (c) Induction motor
    (d) None of the above

Answers

1. (b)	2. (a)	3. (c)	4. (b)	5. (c)
6. (d)	7. (c)	8. (c)	9. (b)	10. (c)

Exercise

1. Describe the PWM control of a three-phase induction motor in detail. Draw its waveform, speed-torque characteristics and obtain the expression of maximum torque. Also, suggest the closed loop scheme with block diagram.

2. Write short notes on variable frequency control of induction motor by VSI and CSI schemes.

3. Variable frequency control of induction motor is more efficient than stator voltage control. Why?

4. Discuss the V/F scheme of speed control of induction motor fed from a voltage source inverter for above and below base speed control. Also, draw the speed-torque characteristics.

5. Write a short note on PWM-controlled induction motor drive.

6. Give reasons for the following:

   (i) Stator voltage control is suitable for speed control of induction motor in fan and pump drives.

   (ii) Stator voltage control is an inefficient method of speed control.

7. Explain variable voltage variable frequency control of an induction motor giving its speed-torque characteristics under different modes of operation.

8. Compare VSI and CSI fed induction motor drives.

9. Starting from the fundamental, prove that torque developed by the induction motor is proportional to the square of the supply voltage.

10. Write a short note on stator voltage control method.

11. A Y-connected SCIM has the following rating and parameters:

    400 V, 50 Hz, 4-pole, 1370 rpm, $R_s = 2$ ohm, $R'_r = 3$ ohm, $X_s = X'_r = 3.5$ ohm motor is controlled by a VSI at constant V/F ratio. The inverter allows frequency variation from 10 to 50 Hz.

    (i) Obtain a pole between the breakdown torque and frequency.

    (ii) Calculate starting torque and current of this drive as a ratio of their values when the motor is started at rated voltage and frequency.

12. A 440 V, 50 Hz, 960 rpm, 6-pole, Y-connected, three-phase slip ring induction motor has the following parameters referred to the stator.

    $R_s = 0.1$ ohm, $R'_r = 0.08$ ohm, $X_s = 0.3$ ohm, $X'_r = 0.4$ ohm.

    The stator to rotor turns ratio is 2.

    The motor is controlled by the static-Scherbius drive. The drive is designed for a speed range of 25% below the synchronous speed. The maximum value of firing angle is 165°. Calculate:

    (i) Transformer turns ratio

    (ii) Torque for a speed of 780 rpm and $\alpha = 150°$

    (iii) Firing angle for half rated motor torque and speed of 800 rpm
       DC link inductor has a resistance of 0.02 ohm.

13. An inverter supplies a 6-pole, three-phase cage induction motor rated at 415 V, 50 Hz. Determine the approximate voltage required of the inverter for motor speed 600/800/1500/1800 rpm.

14. A 440 V, 50 Hz, 6-pole, star connected wound-rotor induction motor has the following parameters referred to stator:

    R_s = 0.5 ohm, R'_r = 0.4 ohm

    X_s = X_r =1.2 ohm, X_m = 50 ohm

    Stator to rotor turn ratio is 3.5. Motor speed is controlled by static rotor resistance control. External resistance is chosen such that the breakdown torque is produced at a standstill for a duty ratio of zero. Calculate the value of external resistance.

15. A 2.8 kW, 400 V. 50 Hz, 4-pole, 1370 rpm, delta connected squirrel-cage induction motor has the following parameters referred to the stator:

    R_s = 2 ohm, R'_s = 5 ohm, X_S = X'_r = 5 ohm, X_m = 80 ohm.

    Motor speed is controlled by stator voltage control. When driving a fan load, it runs at rated speed at rated voltage. Calculate:

    (i) Motor terminal voltage, current

    (ii) Torque at 1200 rpm.

16. A 3-phase, 400 V, 15 kW, 1440 rpm, 50 Hz, star connected induction motor has rotor leakage impedance of 0.4 + j 1.6 ohm. Stator leakage impedance and rotational losses are assumed negligible. If this motor is energized from 120 Hz, 400 V, three-phase source, then calculate:

    (i) Motor speed at rated load

    (ii) Slip at maximum torque

    (iii) Maximum torque

7

Rotor Control of Induction Motor

7.1 INTRODUCTION

Nowadays, improvement for both the output crest voltage and the harmonic copper loss are demonstrated by space vector theory. The electric torque of an induction motor can be explained by the interaction between the rotor currents and the flux resulting from the stator currents. The current is replaced by an equivalent quantity, as the rotor current cannot be measured with squirrel-cage motors. The dynamic model of the induction motor is necessary to understand and analyze vector control theory. It has been found that the dynamic model equations developed on a rotating reference frame is easier to describe the characteristics of the induction motor. The rotor resistance control method can provide high starting torque with low starting current and variation of speed over a wide range below the synchronous speed of the motor. Moreover, the power factor is generally improved. Thus, it is extensively used where a starting current may cause serious line disturbances when the simplicity of operation is desired. Resistance controllers in the rotor circuit of an induction motor are used to achieve smooth start and speed control. In a variable speed drive, the analysis can be conveniently achieved in terms of a mathematical model utilizing the representation of the machine in the synchronously rotating reference frame, whereby the direct and quadrature axes are rotating at synchronous speed.

The dynamic model considers the instantaneous effects of varying voltages/currents, stator frequency, and torque disturbance. The dynamic model of the slip ring induction motor is derived by using a two-phase motor in direct (d) and quadrature (q) axes. This method is easy because of the conceptual simplicity obtained with two sets of windings, one on the stator and the other on the rotor. According to the concept of power invariance the power must be equal in the three-phase machine and its equivalent two-phase model. The differential equations describing the induction motor are non-linear. The dynamic performance of an AC machine is quite complex because of the coupling effect between stator and rotor phases where the coupling coefficients vary with the rotor position.

7.2 STATIC ROTOR RESISTANCE CONTROL

Power electronic devices have shown their performance in controlling the speed of induction motors by varying the resistance at the rotor end. That respective connection diagram has been shown in Fig. 7.2 From the diagram it is clear that the output voltage is rectified by diode connection, which gives the DC output voltage. Further, there is a parallel combination of a transistor T_r and fixed resistance R. Effective value of resistance can be varied by varying the value of duty ratio of the transistor T_r. For ripple, the free output from rectifier and continuity in the DC link inductance L_d is connected between the diode and transistor.

The torque-slip curves of an induction motor are shown in Fig. 7.1 for different values of resistances. As R_r increases, the curve becomes flatter that leads to a lower speed until the motor halts.

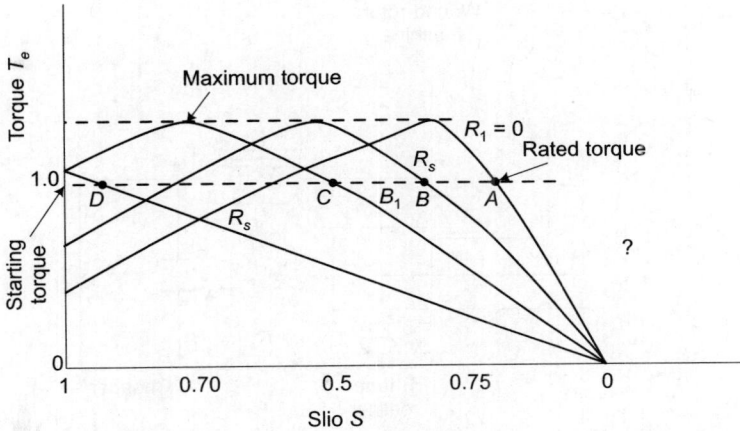

Fig. 7.1 Torque-slip curve of rotor resistance control.

The torque-slip equation for an induction motor is given by

$$T_e = 3\left(\frac{P}{2}\right)\frac{R_r}{s\,\omega_e} \cdot \frac{V_s^2}{(R_s + R_r/s)^2 + \omega_e^2\,(L_{ls} + L_{lr})^2}$$

Though the strategy is quite simple but energy at resistance is wasted. This energy is called *slip energy wastage*.

It is clear that the torque-slip curves are dependent on the rotor resistance R_r. The curves for different rotor resistances are shown on the next slide for four different rotor resistances $(R_1 - R_4)$ with $R_4 > R_3 > R_2 > R_1$ With $R_1 = 0$, i.e., slip rings shorted, speed is determined by rated load torque. As R_r increases, the curve becomes flatter leading to a lower speed until speed becomes zero for $R_r > R_4$.

Advantages of rotor resistance control

- Smooth operation
- Cost effective in comparison to variable frequency control method
- Easy maintenance
- Compact size
- Simple control loop

Disadvantages of rotor resistance control

- Less efficient due to slip energy wastage

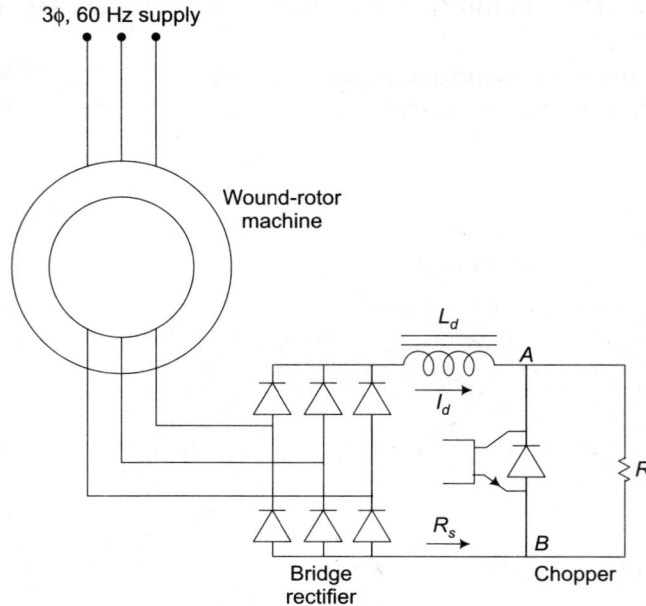

Fig. 7.2 Rotor resistance control of induction motor.

This method is applicable in cranes, Ward Leonard drives, and other short duration load equalization.

Applications of rotor resistance control
• It is employed in cranes, Ward Leonard rives, and other intermittent load applications.

Mathematical equations

Thus, the rms rotor current will be

$$I_r = \sqrt{2/3}\, I_d$$

and resistance between the terminals A and B will be zero if the transistor is on and it will be R if the transistor is off. Therefore, average resistance between the terminals is given by:

$R_{ab} = (1 - \delta)\, R$ where, δ is duty ratio of the transistor.

Power consumed by R_{ab} is $P_{ab} = I_d^2\, R_{ab} = I_d^2\, R\, (1 - \delta)$

From equations 1 and 2, power consumed by R_{ab} per phase is

Power consumed per phase $= \dfrac{p}{3} = 0.5\, R\, (1 - \delta)\, I_r^2$

It is very much clear that the rotor circuit resistance is increased by value $0.5\, R\, (1 - \delta)$. Thus, total circuit resistance per phase will now be R_{rr} which can be varied from R_r to $(R_r + 0.5\, R)$ as δ is changed from 1 to 0.

A closed loop speed control scheme consisting of the current control loop is shown in Fig. 7.2. Rotor current I_r and, therefore, rectified current I_d remains constant at maximum torque point, both during motoring and generating. The current limiter is present in the closed current control loop, which results in acceleration and deceleration about the maximum torque.

This results in very fast transient response. For plugging to occur, arrangement for reversal of phase sequence should be framed.

7.3   STATIC KRAMER DRIVE

To recover the slip energy wastage, the approach has been made to feed this waste back to the AC line by converting it into an appropriate signal. Two methods in support of this statement are:

1. **Static Kramer Drive:** A static Kramer drive is a method to obtain an injected voltage that is in phase with the rotor current.

2. **Static Scherbius Drive:** Static Scherbius drives are capable of bidirectional power flow, with both positive and negative, injected voltages possible, in phase with or opposing the rotor current.

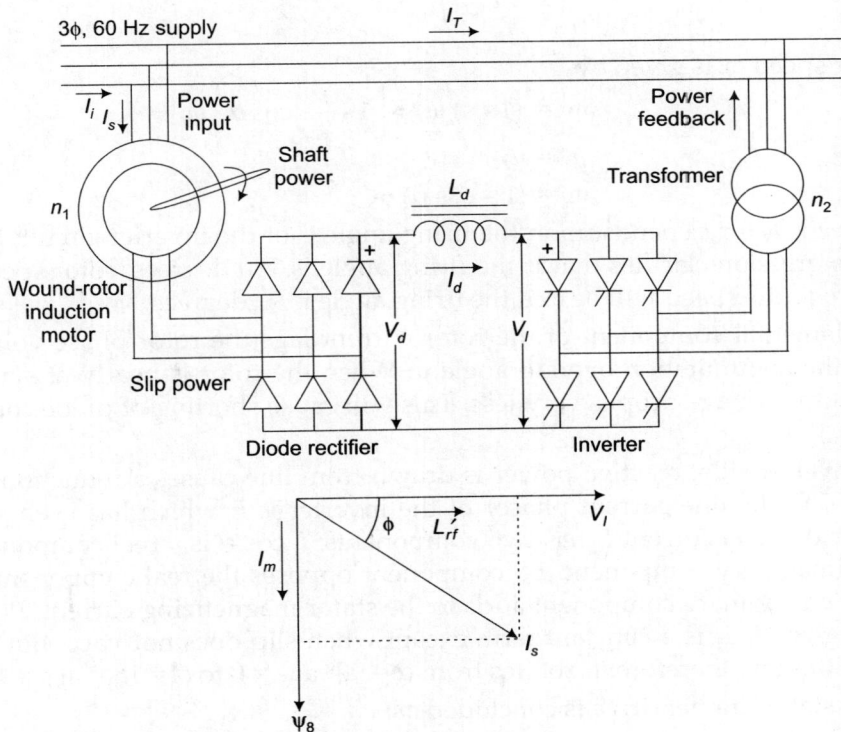

Fig. 7.3   Static Kramer drive.

The voltage V_d is proportional to slip s, and the current I_d is proportional to torque. At a particular speed, inverter firing angle can be varied to vary V_1. This will proportionally vary with the rectified current and, therefore, the torque. The voltage V_d is given by:

$$V_d = \frac{1.35\, s\, V_L}{n_1}$$

where

$$s = \text{slip}$$
$$V_L = \text{Stator line voltage}$$
$$n_1 = \text{Stator to rotor turns ratio}$$

The inverter input voltage is given by:

$$V_1 = \frac{1.35\, s\, V_L}{n_2} \cos \alpha$$

where

$$n_2 = \text{Transformer turns ratio}$$
$$a = \text{Inverter firing angle}$$

In steady state $V_d = V_1$

Therefore,

$$s = \frac{V_1}{V_2} \cos \alpha$$

The rotor speed ω_r is given by:

$$\omega_r = (1 - s)\,\omega_e = \left(1 - \frac{n_1}{n_2} \cos \alpha\right)\omega_e$$

If,

$$n_1 = n_2$$
$$\omega_r = (1 - \cos \alpha)\,\omega_r$$

Thus, speed factor depends upon the firing angle α of the inverter circuit. Moreover, the above expression clarifies that if the firing angle is 180 degrees, rotor speed will be equal to zero while speed will be ω_e if the firing angle is 90 degrees.

The fundamental component of the rotor current lags the rotor phase voltage by φ_r because of the commutation overlap angle μ. When the rotor is nearly at synchronous speed, the rotor voltage drop is very less. This will cause shorting of diodes of the same arm.

On the inverter side, reactive power is drawn from line causes a reduction in power factor ($\varphi_L > \varphi_s$). The line current phasor of the inverter is I_T which has been shown for $n_1 = n_2$ at $s = 0.5$. The current I_T has two components: $I_T \cos \alpha$ is a real component and $I_T \sin \alpha$ is an imaginary component. E_{aL} component opposes the real component of stator current while imaginary component adds to the stator magnetizing current. Phasor sum of I_T and I_S gives I_L. I_T is a constant term except when slip does not vary. But this is no longer the situation, therefore, I_T rotates from $\alpha = 90°$ at $s = 0$ to $\alpha = 180°$ at $s = 1$.

Thus, the static Kramer drive is concluded as:

• At zero speed, the motor is merely a transformer. Therefore, all the real power is fed back to the line. Also, only reactive power is consumed by the motor and inverter.

• At synchronous speed, the motor is in a retarded position, where power factor is minimum. It will increase only if slip increases. Using of step-down transformer shows the improvement in power factor. However, the conflicting point is that the reduced transformer turns ratio reduces the power factor.

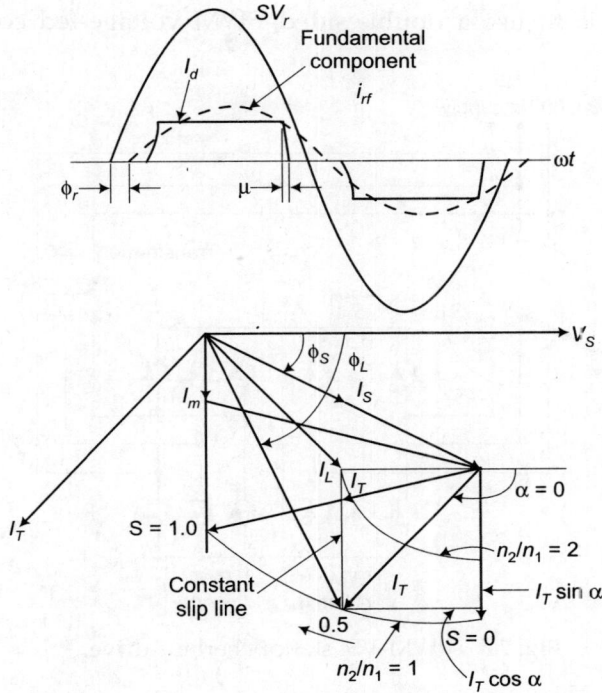

Fig. 7.4 Fundamental component of rotor current and phasor diagram of rated voltage of static Kramer drive.

7.4 STATIC SCHERBIUS DRIVE

Overcoming the only limitation of Kramer drive that it can provide only forward motoring, Scherbius drive is presented with the approach of regenerative mode operation; that is based on the approach of slip power flow in reverse direction. This concept is achieved by replacing the diode converter with the thyristor converter.

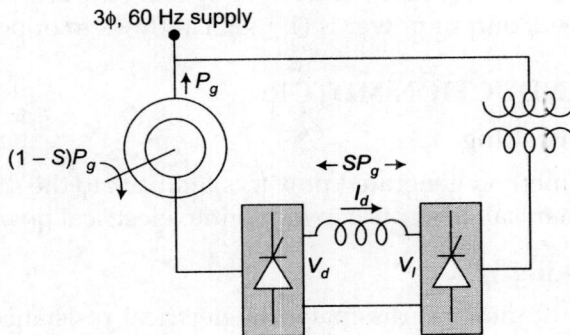

Fig. 7.5 DC link thyristor converter static Scherbius drive.

Another approach is to use a double-sided PWM voltage-fed converter system as shown below:

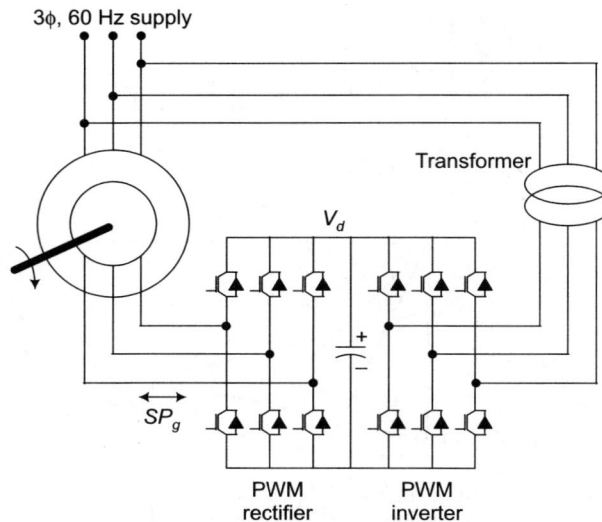

Fig. 7.6   PWM-VSI static Scherbius drive.

This is the basic difference between the two drives.

Another limitation of Kramer drive is that line commutation of machine side converter is difficult near synchronous speed because of excessive commutation angle overlap. However, line commutated cycloconverter overcomes this limitation on the cost and complexity. However, there is even substitute to this problem in which double-sided PWM voltage fed converter system is used as shown in Fig. 7.6.

The resultant air gap flux rotates at synchronous speed when the slip frequency and its phase sequence are adjusted for varying shaft speed.

Slip power is given by sP_m. At subsynchronous speed, output power is $(1 - sP_m)$ and at supersynchronous speed, output power is $(1 + sP_m)$ at flows in opposite direction.

7.5   BRAKING IN INDUCTION MOTOR

7.5.1   Regenerative Braking

Regenerative: In this method generated power is fed back to the source. Motor acts like a machine where mechanical power is converted into electrical power.

7.5.2   Dynamic Braking

The generated energy of motor is dissipated in electrical resistances for slowing down the motor.

7.5.3   Plugging or Plug Reversal

Plugging occurs by interchanging phase sequence connection. Thus, operation shifts from motoring to braking.

7.6   SPECIAL ELECTRICAL DRIVES

In special electrical drives, switched reluctance motor has an important role, it has been receiving attention for commercial and domestic applications due to its high-efficiency simplicity, low cost, and fault tolerance. Rotor position sensors are necessary in SRM in order to synchronize the phase excitation to the rotor position. This has led to attempts to find an alternative way to detect rotor position such as indirect rotor position sensing. Various methods of sensorless position estimation have been investigated for switched reluctance machines. The active phase current detection method using the PWM voltage control, the impedance sensing method, the monitoring current waveforms method, the state observer method, the flux-current detection method, the mutually induced voltage method, the back electromotive force (EMF) method, the capacitive sensing method, the combining opposite-connect sensing coils method. The main idea behind all of these methods is to utilize SRM's salient structure. The magnetic status of the SRM is a function of its rotor position. Therefore, one can acquire the position information that is stored in the magnetic characteristic.

In some applications, such as an electrical vehicle, electric bicycle, reliability, size and costs are design criteria. A simple sensorless control method with no extra hardware and high-performance digital signal processor required is attractive. Most of the proposed sensorless methods require additional hardware or large lookup tables to store magnetic characteristics. The sensorless control scheme requires neither extra hardware nor huge memory space for implementation. And it can be implemented in a low-cost microcomputer.

In order to achieve the required voltage, the photovoltaic (PV) module may be connected in parallel or series, but it is costlier. Thus, to make it cost effective, power converters and batteries are used. The electrical charge is consolidated from the PV panel and directed to the output terminals to produce a low DC voltage. The charge controllers direct this power acquired from the solar panel to the batteries. According to the state of the battery, the charging is done, so as to avoid overcharging and deep discharge. The voltage is then boosted up using the boost power converter, ultimately running the BLDC motor which is used as the drive motor for our vehicle application. The characteristic features of the components; solar panel, charge controller, battery, power converter and BLDC motor drive required for the vehicle application are in real time.

7.6.1   Traction Drives

A traction drive consists of one or more electrical machines and power converters linked by an appropriate control system. The input to the drive is raw AC or DC electrical power from the traction line. The purpose of the drive is to move trains or vehicles according to the kinematic and dynamic requirements of the traction duty cycle, which is expressed as mechanical power and torque as functions of time. The drive output is, therefore, a profile of train motion explained as velocity and acceleration as functions of distance along the track of train.

7.6.1.1   *Working of traction drives*

Ideally, the characteristics of DC series are suitable for electric traction motor. Fast acceleration is the result of high starting torque. In past, diode rectifier system fed these motors. The secondary of a step-down transformer fed through rectifier system, whereas the high voltage AC power system is connected to the primary. Voltage control is provided by transformer tapping. Presently, DC traction motors are consistently fed through converter. Since the converter's output voltage can be controlled by firing angle control, there is no need of transformer tappings. Accurate control and fast response are provided by thyristorised control. Therefore, separately excited DC motors are used by such systems.

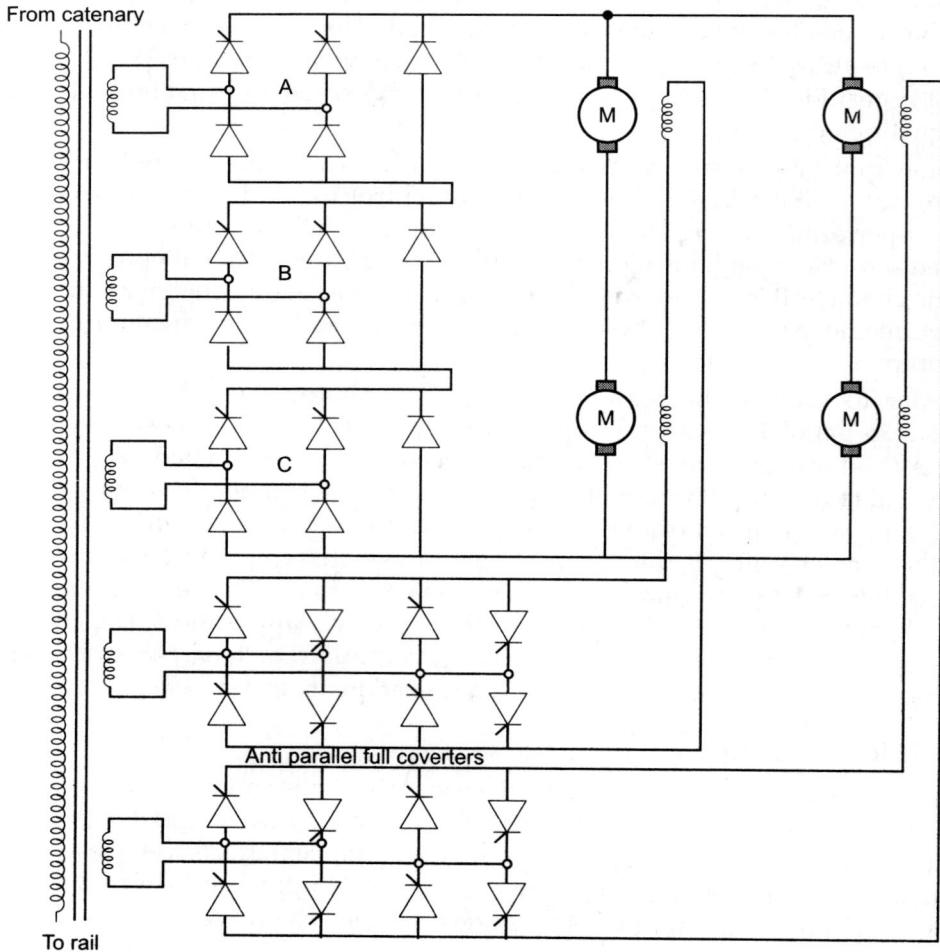

Fig. 7.7   Electrical configuration of traction drive.

Figure 7.7 shows a group of four separately excited motors are fed by thyristorised DC electric system. Half controlled bridge converters are fed through the armature. To reduce the ripple in field current, it is desirable to feed the field winding through fully controlled bridge converters.

In the machine, low iron losses are confirmed by a low ripple in the field current. However, if regenerative braking is required then fully controlled bridges should feed armatures. To ensure the good armature current waveform, freewheeling diodes are connected as shown in the figure. To ensure the good starting and running characteristics, series-parallel arrangement is used to connected armature. It can be seen from the figure that three bridges connected in series are used to feed the armature. For starting first only bridge A is triggered and as speed builds up the firing angle is increased up to saturation level. Bridge B and bridge C are triggered when bridge A is fully conducting (i.e., $\alpha = 0$). While starting to ensure the high starting torque field current are set to maximum.

7.6.2   Diesel-electric Drives

Diesel-electric drives convert movable mechanical power from a diesel engine to traction mechanical power at the train wheels using electrical transmission and motor drives. The diesel engine is coupled to a generator, which produces current that is then conditioned to a form in which it can drive one or more electric traction motors. Diesel-electric drives are used in shunting, trip freight and main line passenger and freight locomotives.

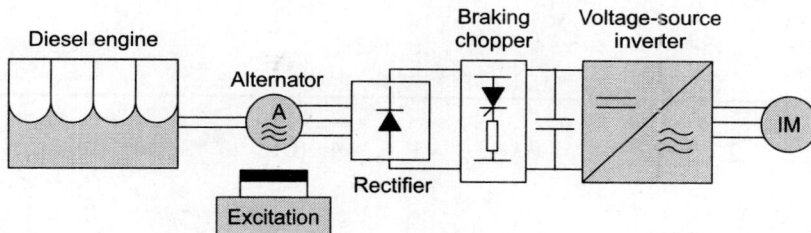

Fig. 7.8   Diesel-electric drives with DC and AC traction motors.

A diesel-electric drive with DC and AC traction motor is shown in Fig. 7.8. The drives with DC traction motors utilize series, mixed excitation or separately-excited motors. Series motors have the advantage of simplicity in their excitation arrangements but do not normally provide dynamic braking down to a standstill. In diesel-electric drive, the diesel engine is coupled to a DC generator which is coupled back-to-back to DC series motors in a Ward-Leonard configuration.

Multiple-Choice Questions

1. Which of the following motors always starts on load
   (a) Conveyor motor
   (b) Floor mill motor
   (c) Fan motor
   (d) All of the these

2. Effect of friction torque is more pronounced
    (a) When the drive is running on full speed
    (b) When the drive is being started
    (c) When the drive is being stopped
    (d) When the drive at half of its normal speed
3. Rotor resistance speed control is used in:
    (a) Squirrel-cage induction motor   (b) Synchronous motor
    (c) Slip ring induction motor   (d) DC shunt motor
4. The cooling time constant is usually:
    (a) Equal to the heating time constant
    (b) More than heating time constant
    (c) Both (a) and (b)
    (d) None of these
5. The motor normally used for crane travel is
    (a) AC slip ring motor
    (b) Ward Leonard controlled DC shunt motor
    (c) Synchronous motor
    (d) DC differentially compound motor

Answers

1. (d)   2. (b)   3. (c)   4. (b)   5. (a)

Exercise

1. Discuss using a block diagram the operation of a voltage source inverter fed synchronous motor in the true synchronous mode.
2. Describe how the speed of a current source inverter fed synchronous motor is controlled in its self-controlled mode. Built a closed loop algorithm to regulate the speed in the above scheme.
3. Describe the self-control of synchronous motor fed from VSI. Discuss the separately controlled synchronous motor fed on VSI. Compare the above two schemes.
4. Show that the torque of a synchronous motor is independent of speed, when it operates in the controlled current mode.
5. Explain the control characteristics of an open loop V/F controlled synchronous motor.
6. Describe in detail, the function of self-controlled synchronous motor employing cycloconverter.

7. Describe using a schematic how the speed of an inverter fed brushless DC motor can be controlled. Bring out its advantages and disadvantages

8. Explain using a diagram the working of a wound field synchronous motor drive with brushless excitation.

9. Explain using a power circuit, the working of a trapezoically excited permanent magnet synchronous motor, operating in the self-controlled mode.

10. Explain how three-phase synchronous motor fed by a three-phase inverter can be made to behave like a simple DC motor. Hence, is it proper to call them commutator less DC motor.

11. Write brief notes on brushless excitation.

12. Write short notes on the following:
    (i) Power factor control of the synchronous motor drive.
    (ii) True synchronous mode of operation.

13. Explain the closed loop control scheme of adjustable speed synchronous motor drive and mention its need and advantages.

14. Suggest a scheme for speed control of synchronous motors using current source inverters and explain its working.

15. Explain the merits and demerits of load commutated CSI.

16. Write short notes on regenerative braking in a synchronous motor.

17. Draw and explain a closed loop operation for a static Kramer controlled drive.

18. In which way, static Kramer control is different from static Scherbius drive?

Numerical Problems

1. A 6 MW 3-phase 11 kV, star connected 6 poles 50 Hz 0.9 lagging power factor synchronous motor has synchronous reactance equal to $9\,\Omega$ and armature resistance equal to $0\,\Omega$. The rated field current is 50 A. The machine is controlled by variable frequency control at constant V/F ratio up to the base speed and at a constant voltage above the base speed. Determine:
   (i) Torque and field current for the rated armature current, 750 rpm and 0.8 leading power factor
   (ii) Armature current and power factor for half the rated motor torque, 1500 rpm and rated field current.

2. A 7 MW 3-phase 12 kV star connected 6-pole 50 Hz 0.9 leading power factor synchronous motor has $X_s = 10\,\Omega$, $R_s = 0\,\Omega$. The rated field current is 40 A. The machine is controlled by variable frequency control at constant V/F ratio up to the base speed and at a constant voltage above base speed. Determine:
   (i) Torque
   (ii) The field current for the rated armature current at 750 rpm and 0.8 leading power factor.

3. A 3-phase 400 volt 50 Hz 6-pole star connected wound rotor synchronous motor has $Z_s = 0 + j2\ \Omega$. Load torque proportional to speed^2 is 340 Nm at rated synchronous speed. The speed of the motor is lowered by keeping V/F constant maintain unity power factor by field control of the motor. For the motor operation at 600 rpm, calculate:

    (i) Supply voltage   (ii) Armature current

    (iii) Excitation angle   (iv) Load angle

    (v) The pullout torque. (Neglect rotational losses.)

8

Indirect Vector Control of Induction Motor

8.1   INTRODUCTION

Electric drives have evolved over the years and so have the techniques to control their speed and torque. There is a large number of research going on to achieve stable control techniques with growing capabilities in related fields; the control techniques are also getting better with time.

The induction motor is widely used in industry because of simple construction, high reliability low cost, etc. In an industrial country, approximately 70% of all generated electrical energy is used by the electrical motors. However, a drawback of the induction motor is that a precise torque control cannot be easily achieved. So, more emphasis is given to find out the mean of precise speed and torque control of induction motor.

These problems can be solved by vector control or field oriented control. The invention of vector control at the beginning of the 1970s, and the demonstration that an induction motor can be controlled like a separately excited DC motor, brought a renaissance in the high-performance control of AC drives. Because of a DC machine like performance, vector control is known as decoupling, orthogonal, or transvector control. Vector control is applicable to both induction and synchronous motor drives.

8.2   PRINCIPLE OF VECTOR CONTROL

Blaschke in 1972 introduced the principle of vector control of induction motor, to realize DC motor characteristics in an induction motor drive. In vector control, both flux and torque are controlled independently, i.e., both are decoupled in nature. The torque equation of DC motor is expressed as:

$$T_e = K_t \, \Psi_f \, \Psi_a = K'_t \, I_a \, I_f \qquad (8.1)$$

where,

I_a = armature current

I_f = field current

In a DC machine, the field flux is perpendicular to armature flux — both are orthogonal to each other, no net interaction is produced by these two fluxes as shown in stationary space phasor diagram **(Fig. 8.1)**. This means that the torque is controlled by adjusting the armature current and Ψ_f is not affected; we get a fast transient response.

An AC machine is not simple because of the interaction between the stator and rotor field, whose orientation is not held at 90 degrees but varies with operating conditions.

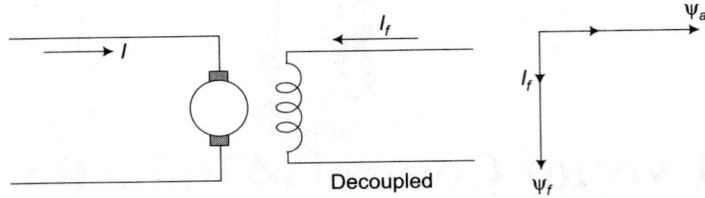

Fig. 8.1   Separately excited DC motor.

We can obtain DC machine-like performance by considering the induction motor in the synchronous rotating reference frame where all sinusoidal variables appear like DC quantity in steady state. The block diagram of vector control implementation with model is shown in (**Fig. 8.2**).

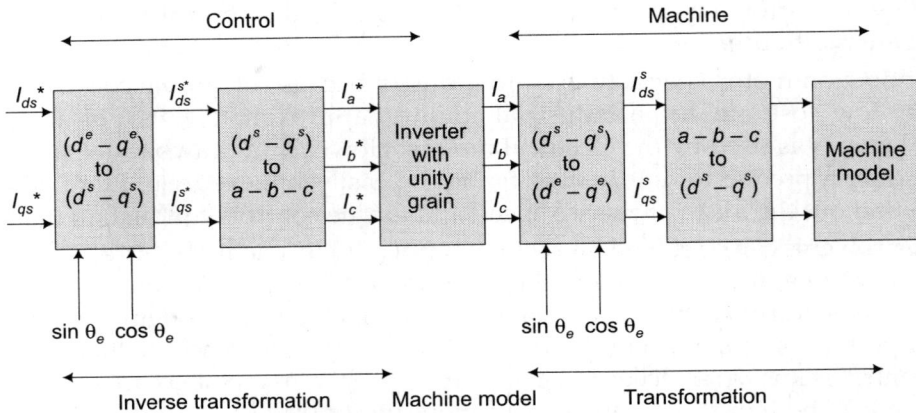

Fig. 8.2   Vector control implementation principle with machine ($d^e - q^e$) model.

The above block diagram is divided in two parts, at the middle inverter with unity current gain is present. The right side of inverter is called machine model transformation in the synchronous rotating frame and the left side is called inverse transformation or control.

The i_a, i_b, i_c are the machine terminal phase currents generated by the inverter and converted to I_{ds}^s & I_{qs}^s i.e., three-phase to two-phase transformation is called Clarke transformation. After that, these I_{ds}^s & I_{qs}^s are again converted to I_{ds} & I_{qs} using unit vector signal, $\sin \theta_e$ & $\cos \theta_e$, is called Park transformation. Then only I_{ds} & I_{qs} is fed to machine model. The torque equation is given in equation 8.2 and phasor diagram of induction motor is shown in **Fig. 8.4**.

$$T_e = K_t \hat{\psi}_r I_{qs} = K_t' I_{qs} I_{ds} \tag{8.2}$$

where, I_{qs} is quadrature axis component of stator current

I_{ds} is direct axis component of stator current

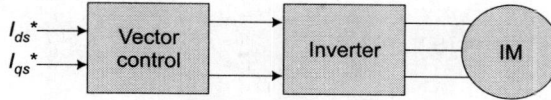

Fig. 8.3   Vector controlled induction motor.

The controller makes two stages of inverse transformation as shown, so that control currents I_{ds}^* & I_{qs}^* correspond to machine current I_{ds} & I_{qs} respectively.

There are two methods of vector control:

1. Direct or feedback method.
2. Indirect or feed forward method.

8.3   DIRECT OR FEEDBACK VECTOR CONTROL

The principal vector controls parameters, i_{ds}^* and i_{qs}^*, which are DC values in the synchronously rotating frame, are converted to stationary frame as vector rotation (VR) with the help of unit vector generated from flux vector signals. The resulting stationary frame signals are then converted to phase current commands for the inverter. The flux signal Ψ_{dr}^s and Ψ_{qr}^s are generated from the machine terminal voltages and currents with the help of a voltage model estimator. A flux control loop has been added for precision control of flux. The torque component of current is i_{qs}^* is generated from the speed control loop through a bipolar limiter. The torque proportional to i_{qs} can be bipolar.

The correct alignment of current in the direction of the flux Ψ_r and the current I_{qs} perpendicular to it are crucial i_{ds} in vector control. This alignment with the help of the stationary frame rotor flux vectors Ψ_{dr}^s & Ψ_{qr}^s. In this **(Fig. 8.4)** $d^e - q^e$ frame is rotating with a synchronous speed ω_e with respect to stationary frame $d^s - q^s$, and at any instant, the angular position of the $d^e - q^e$ to $d^s - q^s$, the axis is θ_e where $\theta_e = \omega_e t$.

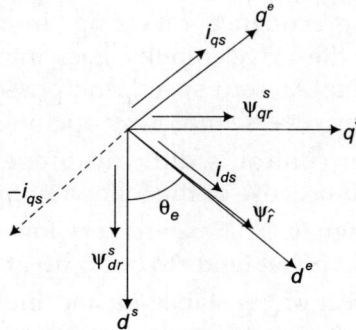

Fig. 8.4 $d^e - q^e$ to $d^s - q^s$ phasors with correct rotor flux orientation.

From Fig. 8.4 we can write the following equations:

$$\psi_{dr}^s = \hat{\psi}_r \cos(\theta_e) \tag{8.3}$$

$$\psi_{qr}^s = \hat{\psi}_r \sin(\theta_e) \tag{8.4}$$

$$\cos(\theta_e) = \frac{\psi_{dr}^s}{\hat{\psi}_r} \qquad (8.5)$$

$$\sin(\theta_e) = \frac{\psi_{qr}^s}{\hat{\psi}_r} \qquad (8.6)$$

$$\bar{\psi}_r = \sqrt{\psi_{dr}^{s2} + \psi_{qr}^{s2}} \qquad (8.7)$$

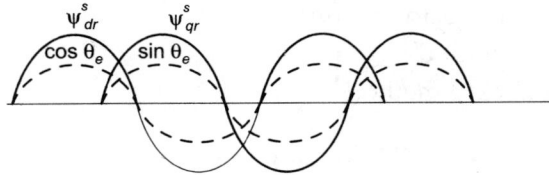

Fig. 8.5   Plot of unit vector signals in correct phase position.

8.4   SALIENT FEATURES OF VECTOR CONTROL

- The frequency ω_e of the drive is not directly controlled as in scalar control. The machine is essentially self-controlled where the frequency, as well as the phase, are controlled indirectly with the help of unit vector.
- There is no fear of an instability problem by crossing the operating point beyond the electromagnetic torque T_{em} as in a scalar control. Limiting the total I_s within the safe limit automatically limits operation within the stable region.
- The transient response will be fast and DC machine like because torque control by i_{qs} does not affect the flux. However, ideal vector control is not possible in practice, because of delays in the converter, signal processing and the variation parameter effect.
- Like a DC machine, speed control is possible in four quadrants without any additional control elements like phase sequence reversing. In forward motoring condition, if the torque T_e is negative, the drive initially goes into regenerative braking mode, which slows down the speed. At zero speed, the phase sequence of the unit vector automatically reverses, giving reverse motoring operation.

The direct method of vector control is difficult to operate successfully at very low frequency including zero speed because of the following problems:

- At low frequency, voltage signals v_{qs}^s & v_{ds}^s are very low. In addition, ideal integration becomes difficult because DC offset tends to build up at the integrator output.
- The parameter variation effect of resistance Rs and inductance L_{ls}, L_{lr} and L_m tend to reduce the accuracy of the estimated signals. Particularly temperature variation of sR becomes more dominant. However, compensation of Rs is somewhat easier, which will be discussed later. At higher voltages, the effect of parameters can be neglected.

In industrial applications, vector drives often require operating from zero speed (including zero speed start-up). Hence, direct vector control with voltage model signal estimation cannot be used.

8.5   INDIRECT (FEED FORWARD) VECTOR CONTROL

The indirect vector control method is essentially the same as direct vector control, except that the unit vector signals are generated in feed-forward manner. Indirect vector control is very popular in industrial applications. The explanation of the fundamental principle of the indirect vector control with the help of phasor diagram is shown in **Fig. 8.6**.

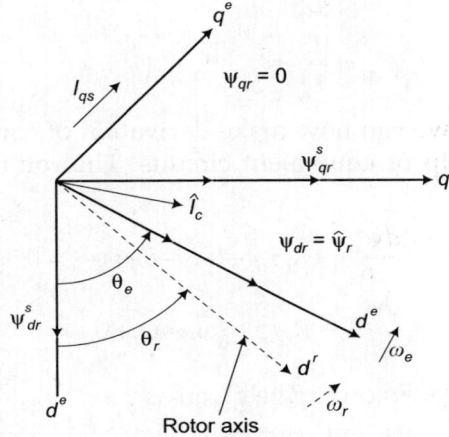

Fig. 8.6   Phasor diagram explaining indirect vector control.

The $d^s - q^s$ axes are fixed on the stator, but the $d^r - q^r$ axes, which are fixed on the rotor, are moving at a speed ω_r as shown in Fig. 8.6. Synchronously rotating axes $d^e - q^e$ are rotating ahead of the $d^r - q^r$ axes by the positive slip angle θ_{sl} corresponding to slip frequency ω_{sl}. Since the rotor pole is directed on the d^e axes and,

$$\omega_e = \omega_r + \omega_{sl} \tag{8.8}$$

We can write

$$\theta_e = \int \omega_e \, dt = \int (\omega_r + \omega_{sl}) \, dt = \theta_r + \theta_{sl} \tag{8.9}$$

The rotor pole position is not absolute but is slipping with respect to the rotor at frequency ω_{sl}. The phasor diagram suggests that for decoupling control, the stator flux component of current i_{ds} should be aligned on the d^e axes, and the torque component of current i_{qs} should be on the q^e axes.

From the matrix equation, the machine terminal voltage and currents are sensed, and fluxes are computed from the stationary frame $d^s - q^s$ equivalent circuit.

These fluxe equations are:

$$\psi_{ds}^s = \int (V_{ds}^s - R_s \, i_{ds}^s) \, dt \tag{8.10}$$

$$\psi_{qs}^s = \int (V_{ds}^s - R_s \, i_{ds}^s) \, dt \tag{8.11}$$

$$\psi_r^{s2} = \psi_{dr}^{s2} + \psi_{qr}^{s2} \tag{8.12}$$

$$\psi_{dim}^s = \psi_{ds}^s - L_{is} \, i_{ds}^s = L_m \, (i_{ds}^s + i_{dr}^s) \tag{8.13}$$

$$\psi_{qim}^s = \psi_{qs}^s - L_{is} \, i_{qs}^s = L_m \, (i_{qs}^s + i_{qr}^s) \tag{8.14}$$

$$\psi_{dr}^s = L_m\, i_{ds}^s + L_r\, i_{dr}^s \tag{8.15}$$

$$\psi_{dr}^s = L_m\, i_{qs}^s + L_r\, i_{qr}^s \tag{8.16}$$

Eliminating i_{dr}^s & i_{qr}^s from equations (8.15) and (8.16) with the help of equations (8.13) and (8.14) respectively, we get,

$$\psi_{dr}^s = \frac{L_r}{L_m}\, \psi_{dm}^s - L_{ir}\, i_{ds}^s \tag{8.17}$$

$$\psi_{qr}^s = \frac{L_r}{L_m}\, \psi_{qm}^s - L_{ir}\, i_{ds}^s \tag{8.18}$$

For decoupling control, we can now make derivation of control equations of indirect vector control with the help of equivalent circuits. The rotor circuit equations can be written as

$$\frac{d\psi_{dr}}{dt} + R_r\, i_{dr} - (\omega_e - \omega_r)\, \psi_{qr} = 0 \tag{8.19}$$

$$\frac{d\psi_{qr}}{dr} + R_r\, i_{qr} - (\omega_e - \omega_r)\, \psi_{dr} = 0 \tag{8.20}$$

The rotor flux linkage expression can be given as

$$\psi_{dr} = L_r\, i_{dr} + L_m\, i_{ds} \tag{8.21}$$

$$\psi_{qr} = L_r\, i_{qr} + L_m\, i_{qs} \tag{8.22}$$

From the above equations, we can write

$$i_{dr} = \frac{1}{L_r}\, \psi_{dr} - \frac{L_m}{L_r}\, i_{ds} \tag{8.23}$$

$$i_{qr} = \frac{1}{L_r}\, \psi_{qr} - \frac{L_m}{L_r}\, i_{qs} \tag{8.24}$$

The rotor currents in equations 8.19 and 8.20 which are accessible can be eliminated with the help of equations 4.23 and 4.24 as,

$$\frac{d\psi_{dr}}{dt} + \frac{R_r}{L_r}\, \psi_{dr} - \frac{L_m}{L_r}\, R_r\, i_{ds} - \omega_{sl}\, \psi_{dr} = 0 \tag{8.25}$$

$$\frac{d\psi_{qr}}{dt} + \frac{R_r}{L_r}\, \psi_{dr} - \frac{L_m}{L_r}\, R_r\, i_{ds} - \omega_{sl}\, \psi_{dr} = 0 \tag{8.26}$$

where

$$\omega = \omega_e - \omega_r \text{ has been substituted.}$$

For decupling control, it is desirable that

$$\psi_{qr} = 0 \tag{8.27}$$

That is,

$$\frac{d\psi_{qr}}{dt} = 0 \tag{8.28}$$

So that, the total rotor flux Ψ_r is directed on the d^e axis.

Substituting the above conditions in equations 8.25 and 8.26, we get

$$\frac{L_r}{R_r} \frac{d\hat{\psi}_r}{dt} + \hat{\psi}_r = L_m \, i_{ds} \tag{8.29}$$

Putting $\tau_r = \dfrac{L_r}{R_r}$ and $\dfrac{d}{dt} = s$ in equation 8.29, we get

$$\tau_r \, s \, \hat{\psi}_r + \hat{\psi}_r = L_m \, i_{ds} \tag{8.30}$$

or

$$\hat{\psi}_r = \frac{L_m \, i_{ds}}{1 + \tau_r \, s} \tag{8.31}$$

From equation 4.26, we get

$$\omega_{sl} = \frac{L_m \, R_r}{\psi_r L_r} \, i_{qs} \tag{8.32}$$

Using the torque equation and putting above condition equation (8.27), we get

$$T_e = \left(\frac{3}{2}\right) \left(\frac{p}{2}\right) \left(\frac{L_m}{L_r}\right) (\psi_{dr} \, i_{qs})$$

or

$$i_{qs} = \left(\frac{2}{3}\right) \left(\frac{2}{p}\right) \left(\frac{L_r}{L_m}\right) \left(\frac{T_e^*}{\hat{\psi}_r}\right) \tag{8.33}$$

To implement the indirect vector control strategy, it is necessary to consider equations 8.9, 8.31, 8.32 and 8.33. The speed control range in indirect vector control can easily be extended from standstill or zero speed to the field weakening region however in field weakening region, the flux is programmed such that the inverter always operates in PWM mode. In both the direct and indirect control methods, instantaneous current control of inverter is necessary. Hysteresis band current control is used for producing switching pulse for the inverter.

8.6 BLOCK OF INDIRECT VECTOR CONTROL OF INDUCTION MOTOR

The closed loop control of motor using indirect vector control is shown in . The control scheme generates inverter switching commands to achieve the desired electromagnetic torque at the shaft of the motor.

8.7 ALGORITHM OF VECTOR CONTROL

1. The induction motor is fed by a current-controlled PWM inverter, which operates as a three-phase sinusoidal current source. The motor speed ω is compared to the reference ω^* and the error is processed by the speed controller to produce a torque command T_e^*.

2. The stator quadrature-axis current reference i_{qs}^* is calculated from torque reference T_e^*

$$i_{qs}^* = \left(\frac{2}{3}\right) \left(\frac{2}{p}\right) \left(\frac{L_r}{L_m}\right) \left(\frac{T_e^*}{\hat{\psi}_r}\right)$$

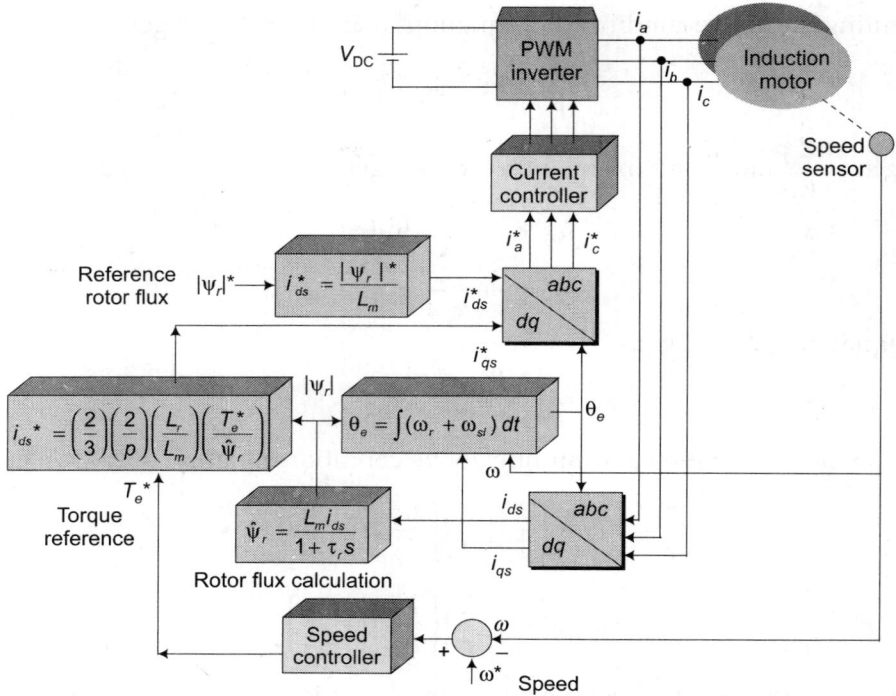

Fig. 8.7   Complete diagram of vector control induction motor.

where $\hat{\psi}_r = |\psi_r|_{est}$ is the estimated rotor flux linkage given by

$$\hat{\psi}_r = \frac{L_m\, i_{ds}}{1 + \tau_r\, s}$$

$w,\, \tau_r = \dfrac{L_r}{R_r}$ is the rotor time constant.

3. The stator direct-axis current reference i_{ds}^* is obtained from reference rotor flux input $|\psi_r|^*$.

$$i_{ds}^* = \frac{|\psi_r|^*}{L_m}$$

4. The rotor flux position θ_e required for coordinates transformation is generated from the rotor speed ω_r and slip frequency ω_{sl} θ_e is given by equation 8.9.

$$\theta_e = \int \omega_e\, dt = \int (\omega_r + \omega_{sl})\, dt = \theta_r + \theta_{sl}$$

5. The slip frequency is calculated from the stator reference current i_{qs}^* and the motor parameters ω_{sl} is given by equation 8.31,

$$\omega_{sl} = \frac{L_m\, R_r}{\psi_r\, L_r}\, i_{qs}$$

6. The i^*_{qs} and i^*_{ds} current references are converted to phase current references i^*_a, i^*_b, i^*_c for the current regulators. The regulators process the measured and reference currents to produce the inverter gating signals. The role of the speed controller is to keep the motor speed equal to the speed reference input in steady state and to provide a good dynamic during transients.

8.8   MATHEMATICAL MODELLING OF SQUIRREL-CAGE INDUCTION MOTOR

In this chapter some reference frame conversions and complex vector notations are used. So, it is quite essential to understand the rest of the theory, it will be shortly described in next subsection.

8.8.1   Clarke Transformation

The Clarke transform is basically employed to transform the three-phase stationary reference frame variable ($a_s - b_s - c_s$) into two-phase stationary reference frame variable ($d^s - q^s$). As shown in Fig. 8.8, ($a_s - b_s - c_s$) axes are displaced at $2\pi/3$ angle and ($d^s - q^s$) axes are oriented at θ angle, where d^s and q^s means stator direct and quadrature axis.

By resolving the ($a_s - b_s - c_s$) into v^s_{ds} & v^s_{qs} component as shown in **Fig. 8.8**.

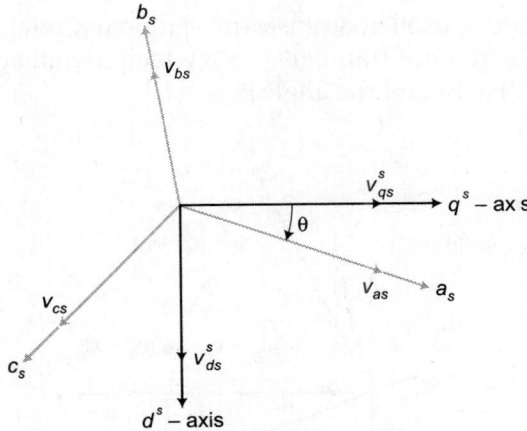

Fig. 8.8   Stationary frame $a - b - c$ to ($d^s - q^s$) transformation.

We get the following matrix:

$$\begin{bmatrix} v^s_{qs} \\ v^s_{ds} \\ v^s_{os} \end{bmatrix} = \frac{2}{3} \begin{bmatrix} \cos\theta & \cos(\theta - 120) & \cos(\theta + 120) \\ \sin\theta & \sin(\theta - 120) & \sin(\theta + 120) \\ 0.5 & 0.5 & 0.5 \end{bmatrix} \begin{bmatrix} v_{as} \\ v_{bs} \\ v_{cs} \end{bmatrix} \tag{8.34}$$

and its inverse is

$$\begin{bmatrix} v_{as} \\ v_{bs} \\ v_{cs} \end{bmatrix} = \begin{bmatrix} \cos\theta & \sin\theta & 1 \\ \cos(\theta - 120) & \sin(\theta - 120) & 1 \\ \cos(\theta + 120) & \sin(\theta + 120) & 1 \end{bmatrix} \begin{bmatrix} v^s_{qs} \\ v^s_{ds} \\ v^s_{os} \end{bmatrix} \tag{8.35}$$

It is convenient to set θ = 0, so that q-axis is coinciding with a_s-axis. The above matrix becomes:

$$
\begin{bmatrix} v_{qs}^s \\ v_{ds}^s \\ v_{os}^s \end{bmatrix} = \frac{2}{3} \begin{bmatrix} 1 & -\dfrac{1}{2} & -\dfrac{1}{2} \\ 0 & -\dfrac{\sqrt{3}}{2} & -\dfrac{\sqrt{3}}{2} \\ 0.5 & 0.5 & 0.5 \end{bmatrix} \begin{bmatrix} v_{as} \\ v_{bs} \\ v_{cs} \end{bmatrix}
\tag{8.36}
$$

and its inverse is:

$$
\begin{bmatrix} v_{as} \\ v_{bs} \\ v_{cs} \end{bmatrix} = \begin{bmatrix} 1 & 0 & 1 \\ -\dfrac{1}{2} & -\dfrac{\sqrt{3}}{2} & 1 \\ -\dfrac{1}{2} & \dfrac{\sqrt{3}}{2} & 1 \end{bmatrix} \begin{bmatrix} v_{qs}^s \\ v_{ds}^s \\ v_{os}^s \end{bmatrix}
\tag{8.37}
$$

8.8.2 Park Transformation

The Park transformation is used to transform stationary reference frame ($d^s - q^s$) to synchronously rotating reference frame ($d^e - q^e$), which is rotated at synchronous speed ω_e with respect to ($d^s - q^s$) axes and the angle $\theta_e = \omega_e t$,

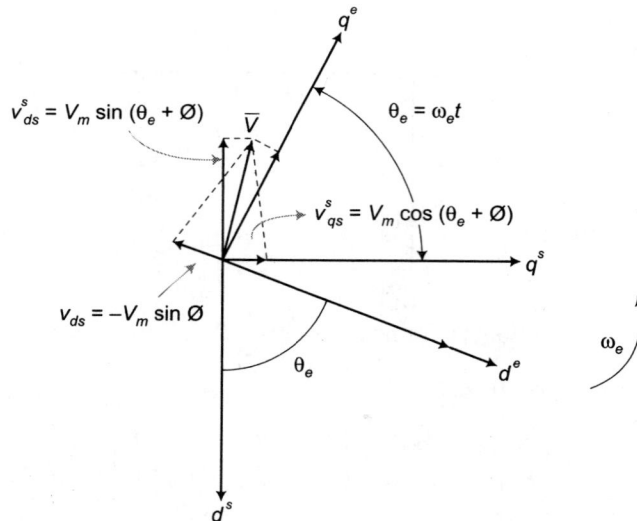

Fig. 8.9   Stationary frame to synchronous frame transformation.

$$
\begin{bmatrix} v_{qs}^e \\ v_{ds}^e \end{bmatrix} = \begin{bmatrix} \cos\theta_e & -\sin\theta_e \\ \sin\theta_e & \cos\theta_e \end{bmatrix} \begin{bmatrix} v_{qs}^s \\ v_{ds}^s \end{bmatrix}
\tag{8.38}
$$

and its inverse is:

$$\begin{bmatrix} v_{qs}^s \\ v_{ds}^s \end{bmatrix} = \begin{bmatrix} \cos \theta_e & \sin \theta_e \\ -\sin \theta_e & \cos \theta_e \end{bmatrix} \begin{bmatrix} v_{qs}^e \\ v_{ds}^e \end{bmatrix} \qquad (8.39)$$

8.8.3 Induction Motor Modelling in Synchronous Rotating Reference Frame and Its Dynamic Model

The three-phase squirrel-cage induction motor is regarded as connected to the 3-wire system. So, rotor winding of motor is equivalent to the 3-wire system.

The line to neutral voltage represented in matrix form is given as:

For stator

For rotor

$$\begin{bmatrix} v_{as} \\ v_{bs} \\ v_{cs} \end{bmatrix} = \begin{bmatrix} i_{as} & \psi_{as} \\ i_{bs} & \psi_{bs} \\ i_{cs} & \psi_{cs} \end{bmatrix} \begin{bmatrix} R_s \\ \dfrac{d}{dt} \end{bmatrix} \qquad \begin{bmatrix} v_{ar} \\ v_{br} \\ v_{cr} \end{bmatrix} = \begin{bmatrix} i_{ar} & \psi_{ar} \\ i_{br} & \psi_{br} \\ i_{cr} & \psi_{cr} \end{bmatrix} \begin{bmatrix} R_r \\ \dfrac{d}{dt} \end{bmatrix} \qquad (8.40)$$

where,

v_{as}, v_{bs}, v_{cs} are stator three-phase supply voltages.

ψ_{as}, ψ_{bs}, ψ_{cs} are flux linkages of phase a, b, c of the stator.

Using Clarke transform, when the above stator and rotor matrices are converted into two-phase, the matrices become:

For stator

For rotor

$$\begin{bmatrix} v_{qs}^s \\ v_{ds}^s \end{bmatrix} = \begin{bmatrix} i_{qs}^s & \psi_{qs}^s \\ i_{ds}^s & \psi_{ds}^s \end{bmatrix} \begin{bmatrix} R_s \\ \dfrac{d}{dt} \end{bmatrix} \qquad \begin{bmatrix} v_{qr}^s \\ v_{dr}^s \end{bmatrix} = \begin{bmatrix} i_{qr}^s & \psi_{qr}^s \\ i_{dr}^s & \psi_{dr}^s \end{bmatrix} \begin{bmatrix} R_r \\ \dfrac{d}{dt} \end{bmatrix} \qquad (8.41)$$

where,

ψ_{qs}^s is q-axes stator flux linkage

ψ_{ds}^s is d-axes stator flux linkage

Using Park transform, stationary ($d^s - q^s$) frame is converted to synchronous rotating frame ($d^e - q^e$), superscript e has been eliminated for convenience. The above matrix becomes:

For stator

For rotor

$$\begin{bmatrix} v_{qs} \\ v_{ds} \end{bmatrix} = \begin{bmatrix} i_{qs} & \psi_{qs} & \psi_{ds} \\ i_{ds} & \psi_{ds} & -\psi_{qs} \end{bmatrix} \begin{bmatrix} R_s \\ \dfrac{d}{dt} \\ \omega_e \end{bmatrix} \qquad \begin{bmatrix} v_{qr} \\ v_{dr} \end{bmatrix} = \begin{bmatrix} i_{qr} & \psi_{qr} & \psi_{ar} \\ i_{dr} & \psi_{dr} & -\psi_{qr} \end{bmatrix} \begin{bmatrix} R_r \\ \dfrac{d}{dt} \\ \omega_e \end{bmatrix} \qquad (8.42)$$

where, all the variables and parameters are referred to the stator. Since, the rotor actually moves at speed ω_r, the d-q axes is fixed on the rotor which moves at speed ($\omega_e - \omega_r$). Therefore, modified rotor matrix becomes:

For rotor

$$\begin{bmatrix} v_{qr} \\ v_{dr} \end{bmatrix} = \begin{bmatrix} i_{qr} & \psi_{qr} & \psi_{dr} \\ i_{dr} & \psi_{dr} & -\psi_{qr} \end{bmatrix} \begin{bmatrix} R_r \\ \dfrac{d}{dt} \\ (\omega_e - \omega_r) \end{bmatrix} \tag{8.43}$$

- **Dynamic Model**

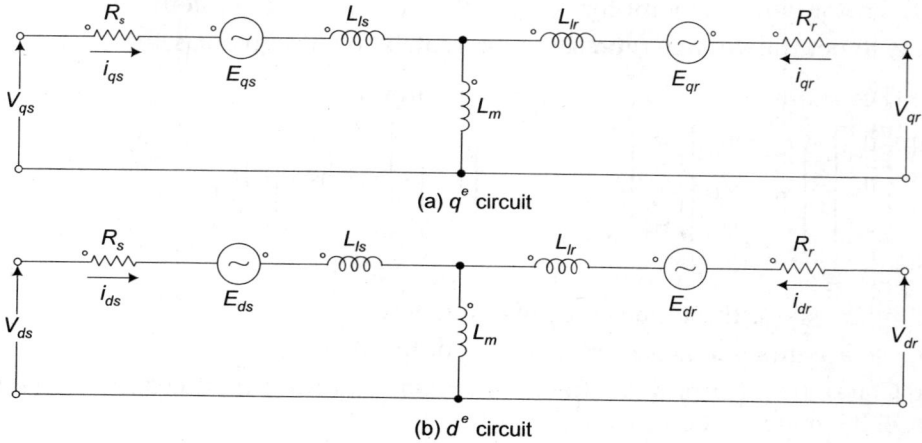

(a) q^e circuit

(b) d^e circuit

Fig. 8.10   (a) and (b): $(d^e - q^e)$ equivalent circuit.

where,

$$E_{qs} = \omega_e \psi_{ds}$$
$$E_{qr} = (\omega_e - \omega_r) \psi_{dr}$$
$$E_{ds} = \omega_e \psi_{qs}$$
$$E_{dr} = (\omega_e - \omega_r) \psi_{qr}$$

Flux linkage expression is given in matrix form as:

$$\begin{bmatrix} \psi_{qs} \\ \psi_{qr} \\ \psi_{qm} \\ \psi_{ds} \\ \psi_{dr} \\ \psi_{dm} \end{bmatrix} = \begin{bmatrix} (L_{ls} + L_m) & L_m & 0 & 0 \\ L_m & (L_{lr} + L_m) & 0 & 0 \\ L_m & L_m & 0 & 0 \\ 0 & 0 & (L_{ls} + L_m) & L_m \\ 0 & 0 & L_m & (L_{lr} + L_m) \\ 0 & 0 & L_m & L_m \end{bmatrix} \begin{bmatrix} i_{qs} \\ i_{qr} \\ i_{ds} \\ i_{dr} \end{bmatrix} \tag{8.44}$$

The instantaneous torque is developed by applying the principle of virtual displacement, Which isgiven by:

$$T_e = \left(\frac{n}{2}\right)\left(\frac{p}{2}\right)(\psi_{qr} i_{dr} - \psi_{dr} i_{qr}) \tag{8.45}$$

where,

n = number of phase,

P = number of poles

Using flux linkage expression, several other torque equations for three phases are given by:

$$T_e = \left(\frac{3}{2}\right)\left(\frac{p}{2}\right)\left(\frac{L_m}{L_r}\right)(\psi_{dr}i_{qs} - \psi_{qr}i_{ds}) \qquad (8.46)$$

$$T_e = \left(\frac{3}{2}\right)\left(\frac{p}{2}\right)(i_{dr}i_{qs} - i_{qr}i_{ds}) \qquad (8.47)$$

8.8.4   Induction Motor Modelling in Stationary Reference Frame (d^s – q^s) and its Dynamic Model

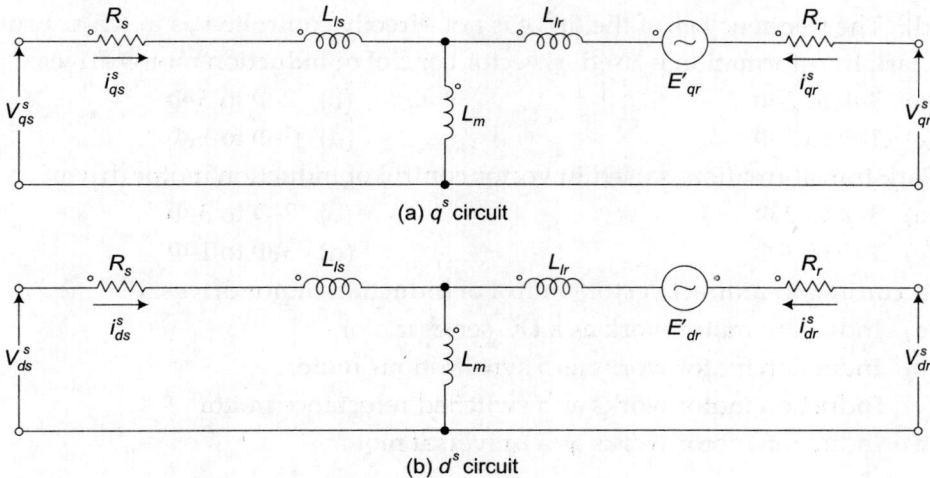

The stator and rotor voltage matrix in stationary frame can be simply derived by putting $\omega_e = 0$ in the above matrix equations. We get:

For stator   For rotor

$$\begin{bmatrix} v_{qs}^s \\ v_{ds}^s \end{bmatrix} = \begin{bmatrix} i_{qs}^s & \psi_{qs}^s \\ i_{ds}^s & \psi_{ds}^s \end{bmatrix}\begin{bmatrix} R_s \\ \dfrac{d}{dt} \end{bmatrix} \qquad \begin{bmatrix} v_{qr}^s \\ v_{dr}^s \end{bmatrix} = \begin{bmatrix} i_{qr}^s & \psi_{qr}^s & -\psi_{dr}^s \\ i_{dr}^s & \psi_{dr}^s & \psi_{qr}^s \end{bmatrix}\begin{bmatrix} R_r \\ \dfrac{d}{dt} \\ \omega_r \end{bmatrix} \qquad (8.48)$$

• Dynamic Model

(a) q^s circuit

(b) d^s circuit

Fig. 8.11   (a) and (b): (d^s – q^s) equivalent circuit.

where,

$$E'_{qs} = \omega_r \psi_{dr}^s$$

$$E'_{dr} = \omega_r \psi_{qr}^s$$

The torque equation in stationary frame is:

$$T_e = \left(\frac{3}{2}\right)\left(\frac{p}{2}\right)(\psi_{dm}^{s}\, i_{qr}^{s} - \psi_{qm}^{s}\, i_{dr}^{s}) \qquad (8.49)$$

$$T_e = \left(\frac{3}{2}\right)\left(\frac{p}{2}\right)(\psi_{dm}^{s}\, i_{qs}^{s} - \psi_{qm}^{s}\, i_{ds}^{s}) \qquad (8.50)$$

$$T_e = \left(\frac{3}{2}\right)\left(\frac{p}{2}\right)(i_{qs}^{s}\, i_{dr}^{s} - i_{ds}^{s}\, i_{qr}^{s}) \qquad (8.51)$$

Multiple-Choice Questions

1. The stator direct-axis current reference is obtained from reference rotor flux input

   (a) $i_{ds} = \dfrac{|\psi_r|}{L_m}$

   (b) $i_{ds} = \dfrac{|\psi_r|^*}{L_m}$

   (c) $i_{ds}^* = \dfrac{|\psi_r|^*}{L_m}$

   (d) $i_{ds}^* = \dfrac{|\psi_r|^*}{L_m + L_{ls}}$

2. Vector control features are:
   (a) The frequency ω_e of the drive is not directly un-controlled as in scalar control
   (b) The frequency ω_e of the drive is directly controlled as in scalar control
   (c) The frequency ω_e of the drive is not directly controlled as in frequency
   (d) The frequency ω_e of the drive is not directly controlled as in scalar control
3. Clark transformation is used in vector control of induction motor drives:
   (a) 3-Φ to 2-Φ
   (b) 2-Φ to 3-Φ
   (c) 1-Φ to 3-Φ
   (d) 3-Φ to 1-Φ
4. Park transformation is used in vector control of induction motor drives:
   (a) 3-Φ to 2-Φ
   (b) 2-Φ to 3-Φ
   (c) 1-Φ to 3-Φ
   (d) 3-Φ to 1-Φ
5. According to indirect vector control of induction motor drives:
   (a) Induction motor work as a DC series motor
   (b) Induction motor works as a synchronous motor
   (c) Induction motor works as a switched reluctance motor
   (d) Induction motor works as a universal motor

Answers

1. (c)   2. (d)   3. (a)   4. (b)   5. (a)

Exercise

1. What is vector control?
2. Describe the working operation of indirect vector control of a three-phase induction motor.
3. State and explain torque relationship of squirrel-cage induction motor.
4. Write comparisons between direct vector control and indirect vector control.
5. What are the applications of vector control of induction motor?
6. Elaborate the salient features of vector control.
7. Draw the phasor diagram of vector control of an induction motor.
8. Write an algorithm for indirect vector control of induction motor.
9. Find the expression for quadrature axis component of current for an induction motor.
10. Write and explain the expression for direct torque and flux for an induction motor.
11. Find the expression for stator and rotor flux control of vector control of induction motor drives.
12. Explain the methods of vector control.
13. Explain the dynamic modelling of induction machine.
14. Write and explain Park's transformation.
15. Write and explain Clark's transformation.
16. Explain the conversion of three-phase AC component to two-phase DC component.
17. Draw the dynamic modelling diagram of the induction motor.

9

Synchronous Motor Drives

9.1   INTRODUCTION

The common feature which links the motors is that they are all AC motors in which the electrical power that is converted to mechanical power is fed into the stator. Hence, in an induction motor, there are no sliding contacts in the main power circuits. All motors except switch reluctance motor have stator identical to the induction motor. These motors are called 'synchronous' or 'reluctance' motors and have a wide range of loads also. They provide a precise and constant speed, so when constant speed operation is required the induction motors are preferred. So from small single-phase versions in domestic timers to multi-megawatt machines in big industrial applications machines are available over a wide range. The principal disadvantage of this motor is that when the load torque becomes high, the motor suddenly loses synchronism and stalls.

In the synchronous motor, the stator windings are exactly the same as in the induction motor, so when connected to a 3-phase supply, a rotating magnetic field is produced. But instead of having a cylindrical rotor with a cage winding, the synchronous motor has a rotor with either a DC excited winding, or permanent magnets, designed to cause the rotor to 'lock-on' or 'synchronize with' the rotating magnetic field produced by the stator. Once the rotor is synchronized, it will run at exactly the same speed as the rotating field despite load variations, so under constant-frequency operation the speed will remain constant as long as the supply frequency is stable.

The armature winding is on the stator and field winding is on the rotor. The three-phase AC supply is connected to the stator (armature winding) and external DC source is given to the rotor field. The speed of rotation is called synchronous speed (Ns)

$$\text{Synchronous speed (Ns)} = \frac{120 \times \text{Frequency}}{\text{No. of Poles}} \text{ rpm}$$

$$\text{Synchronous speed (Ws)} = \frac{4 \times \Pi \times \text{Frequency}}{\text{No. of Poles}} \text{ rad/sec.}$$

Synchronous motor develops torque only at synchronous speed. It is not self-starting.

In a synchronous motor, a 3-phase set of stator currents produces a rotating magnetic field causing the rotor magnetic field to align with it. The rotor magnetic field is produced by a DC current applied to the rotor winding. Field windings produce the main magnetic field (rotor windings for synchronous machines); armature windings are the windings where the main voltage is induced (stator windings for synchronous machines).

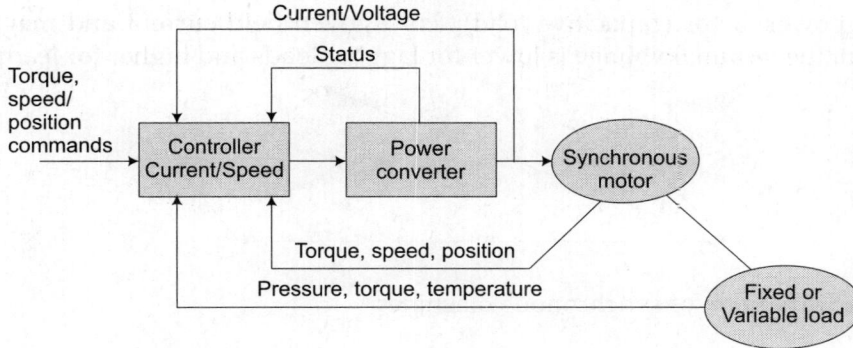

Fig. 9.1   Block diagram of synchronous motor drives.

9.2   OPERATION

Synchronous motor works on the principle of "magnetic locking". When two unlike poles are brought near each other, there exists a tremendous force of attraction between those two poles.

The rotor winding is energized by direct current (DC), and the stator winding is energized by alternating current (AC) of frequency in Hz. There will be average torque produced only when the machine rotates at the same speed as the rotating magnetic field produced by the stator winding.

The synchronous motor is not self-starting due to continuous and rapid rotation of stator poles, the rotor is subjected to a torque which is rapidly reversing, i.e., in quick succession, the rotor is subjected to a torque which tends to move in one direction and then in the other direction. Owing to its large inertia, the rotor cannot instantaneously respond to such quickly-reversing torque, with the result that it remains stationary.

Since the voltages in a synchronous generator are AC voltages, they are usually expressed as phasors. A vector plot of voltages and currents within one phase is called a phasor diagram.

A phasor diagram of a synchronous generator with a unity power factor (resistive load)

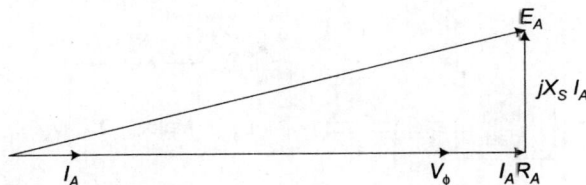

Lagging power factor (inductive load): A larger than for leading PF internal generated voltage E_A is needed to form the same phase voltage.

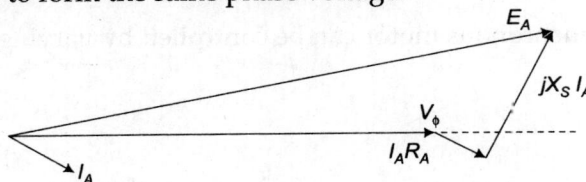

Leading power factor (capacitive load): For a given field current and magnitude of load current, the terminal voltage is lower for lagging loads and higher for leading loads.

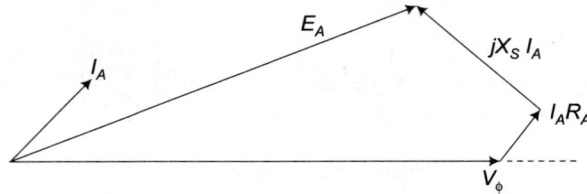

Power flow diagram of synchronous machines

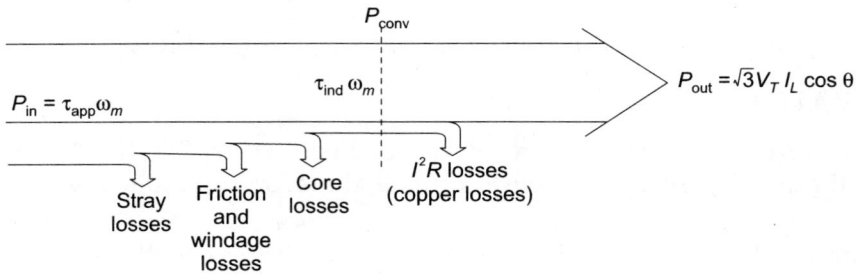

Fig. 9.2   Power flow diagram of synchronous motor.

9.3   SYNCHRONOUS MOTOR DRIVES

A synchronous motor cannot start in synchronous mode since the inertia, and the mechanical load prevent the rotor from catching up with the rotating magnetic field at the synchronous speed. A common practice is to embed a few copper or aluminium bars short-circuited by end rings in the rotor and to start the motor as an induction motor. When the rotor speed is close to the synchronous speed, the rotor is energized with a DC power supply, and it will catch up or synchronize with the rotating magnetic field. The motor drive is suitable for a constant torque load when the speed is below the rated speed, and would be suitable for a constant power load when the speed is higher than the rated speed.

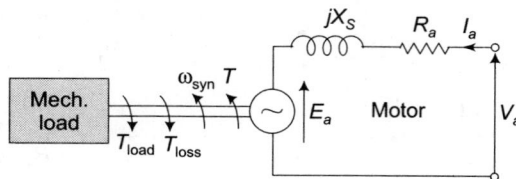

Fig. 9.3   Synchronous motor with mechanical load.

- The speed of the synchronous motor can be controlled by varying the frequency of its source.

- Earlier, synchronous motors were mainly used in constant speed applications, but the development of semiconductor variable frequency sources such as cycloconverters and inverters has allowed their use in variable speed applications.
- Variable speed drives using synchronous motor has high power applications such as pumps, fans, conveyors, main line tractions, compressors, blowers, servodrives, machine tools, textile mills, cement mills, rolling mills, etc.

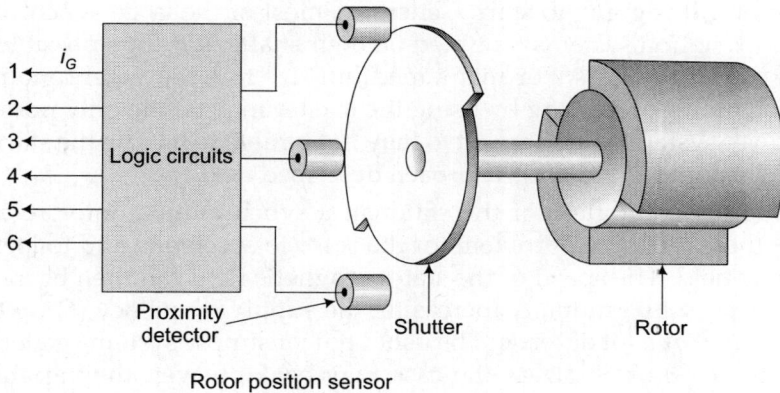

Fig. 9.4   Speed control of synchronous motor.

9.4   STARTING METHODS OF SYNCHRONOUS MACHINE

The various methods to start the synchronous motor are:

1. Using small motors
2. Using external prime mover
3. Using damper winding
4. Using a slip ring induction motor
5. Using a small DC machine coupled with it.

9.4.1   Using Small Motors

This method of starting a synchronous motor is to attach an external starting small motor to it and bring the synchronous machine to near about its rated speed but not exactly equal to it, as the synchronization process may fail to indicate the point of closure of the main switch connecting the synchronous machine to the supply system with the small motor. Then the output of the synchronous machine can be synchronized or paralleled with its power supply system as a generator, and the small motor can be detached from the shaft of the machine or the supply to the small motor can be disconnected. Once the external small motor is turned off, the shaft of the machine slows down the speed of the rotor magnetic field. The synchronous machine continues to operate as a motor. As soon as it begins to operate as a motor the synchronous motor can be loaded in the usual manner just like any motor.

This whole procedure is not as cumbersome as it sounds, since many synchronous motors are parts of motor-generator sets, and the synchronous machine in the motor-generator set may be started with the other machine serving as the starting motor. Moreover, the starting motor is required to overcome only the mechanical inertia of the synchronous machine without any mechanical load (load is attached only after the synchronous machine is paralleled to the power supply system). Since only the motor's inertia must be overcome, the starting motor can have a much smaller rating than the synchronous motor it is going to start. Generally, most of the large synchronous motors have brushless excitation systems mounted on their shafts. It is then possible to use these exciters as the starting motors. For many medium-size to large synchronous motors, an external starting motor or starting by using the exciter may be the only possible solution because the power systems they are tied to may not be able to handle the starting currents needed to use the damper winding approach described next.

If the rotating magnetic field of the stator in a synchronous motor rotates at a low enough speed, there will be no problem for the rotor to accelerate and to lock in with the stator's magnetic field. The speed of the stator magnetic field can then be increased to its rated operating speed by gradually increasing the supply frequency (f) up to its normal 50 or 60 Hz. This makes a lot of sense. The usual power supply systems generally regulate the frequency to be 50 or 60 Hz as the case may be. However, the variable frequency voltage source can be obtained from a dedicated generator only, and such a situation is obviously impractical except for very unusual or special drive applications which can be used to convert a constant frequency AC supply to a variable frequency AC supply. With the development of such modern solid-state variable-frequency drive packages, it is thus

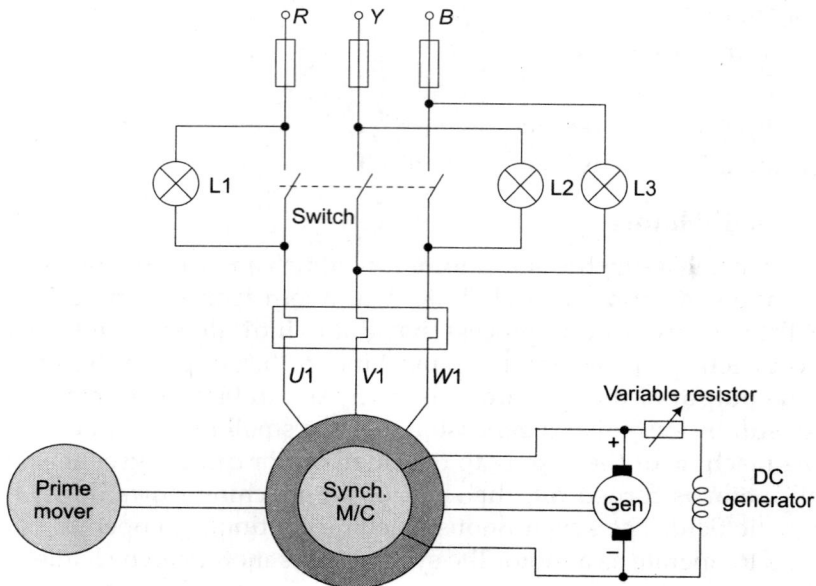

Fig. 9.5   Starting of a synchronous motor using small motor.

possible to continuously control the frequency of the supply connected to the synchronous motor all the way from a fraction of a hertz up to and even above the normal rated frequency. If such a variable-frequency drive unit is included in a motor-control circuit to achieve speed control, then starting the synchronous motor is very easy–simply adjust the frequency to a very low value for starting, and then raise it up to the desired operating frequency for normal running. When a synchronous motor is operated at a speed lower than the rated speed, its internal generated voltage (counter EMF) is smaller than normal. As such the terminal voltage applied to the motor must be reduced proportionally with the frequency in order to keep the stator current within the rated value. Generally, the voltage in any variable-frequency power supply varies roughly linearly with the output frequency.

9.4.2   Using External Prime Mover

To use an external prime mover to accelerate the rotor of the synchronous motor near to its synchronous speed and then supply the rotor as well as a stator. It should be taken to ensure that the direction of rotation of the rotor, as well as that of the rotating magnetic

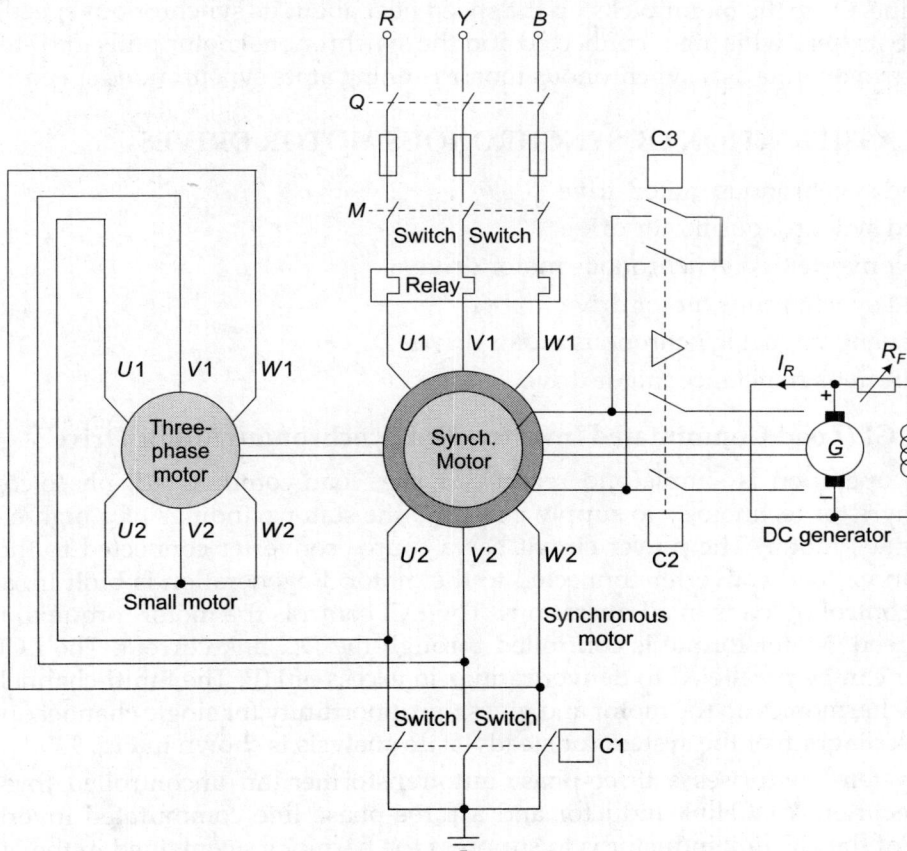

Fig. 9.6   Starting of synchronous motor using external prime mover.

field of the stator, is the same. The synchronous machine is started as a generator and is then connected to the supply mains by following the synchronization or paralleling procedure. Then the power supply to the prime mover is disconnected so that the synchronous machine will continue to operate as a motor.

9.4.3   Using Damper Winding

The damper windings are provided in most of the large synchronous motors in order to nullify the oscillations of the rotor whenever the synchronous machine is subjected to a periodically varying load.

Damper windings are special bars laid into slots cut in the pole face of a synchronous machine and then shorted out at each end by a large shorting ring, similar to the squirrel-cage rotor bars. A pole face with a set of damper windings. When the stator of such a synchronous machine is connected to the 3-phase AC supply, the machine starts as a 3-phase induction machine due to the presence of the damper bars, just like a squirrel-cage induction motor. Just as in the case of a 3-phase squirrel-cage induction motor, the applied voltage must be suitably reduced so as to limit the starting current to the safe rated value. Once the motor picks up the speed near about its synchronous speed, the DC supply to its field winding is connected and the synchronous motor pulls into step, i.e., it continues to operate as a synchronous motor running at its synchronous speed.

9.5   CLASSIFICATION OF SYNCHRONOUS MOTOR DRIVES

- LCI fed synchronous motor drive
- VSI fed synchronous motor drive
- Cycloconverter fed synchronous motor drive
- CSI fed synchronous motor drive
- Permanent magnet synchronous motor drive
- Synchronous reluctance motor drive

9.5.1   LCI (Load Commutated Inverter) Fed Synchronous Motor Drive

The LCI operation is simple and reliable. It uses load-commutated, phase-controlled power thyristor technology to supply power to the stator windings of a high-efficiency synchronous motor. The power circuit has a source converter connected to the power supply and a load converter connected to the motor. Regeneration is built in, allowing smooth control of loads in all operations. The LCI controls the motor torque to regulate motor speed. Motor torque is controlled through the DC link current. The LCI power converter can be paralleled to deliver ratings in excess of HP. The multi-channel output gives low harmonics on the motor and gives the opportunity for single channel operation. The block diagram of the system for steady-state analysis is shown in **Fig. 9.7.**

The system comprises a three-phase autotransformer, an uncontrolled three-phase bridge rectifier, a DC link inductor and a three-phase line commutated inverter. The function of the DC link inductor is to suppress the harmonics contained in the output of the bridge rectifier. The combination of the uncontrolled rectifier and DC link inductor

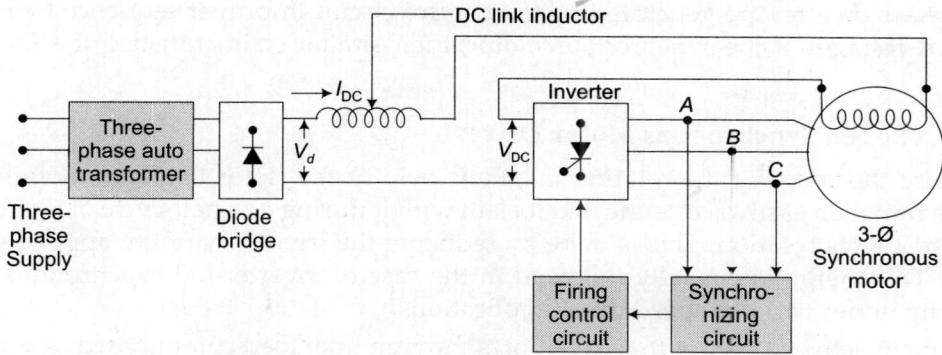

Fig. 9.7 Block diagram of LCI fed synchronous motor drives.

acts as a current source for the inverter. The excitation winding of the synchronous machine is connected in series to the input of the inverter. The synchronizing signal is obtained by sensing line-to-line voltages with the help of a small step-down transformer from the synchronous machine terminals. This synchronizing signal is in input to the control circuit for generating firing pulses for six SCRs of the inverter in proper sequence.

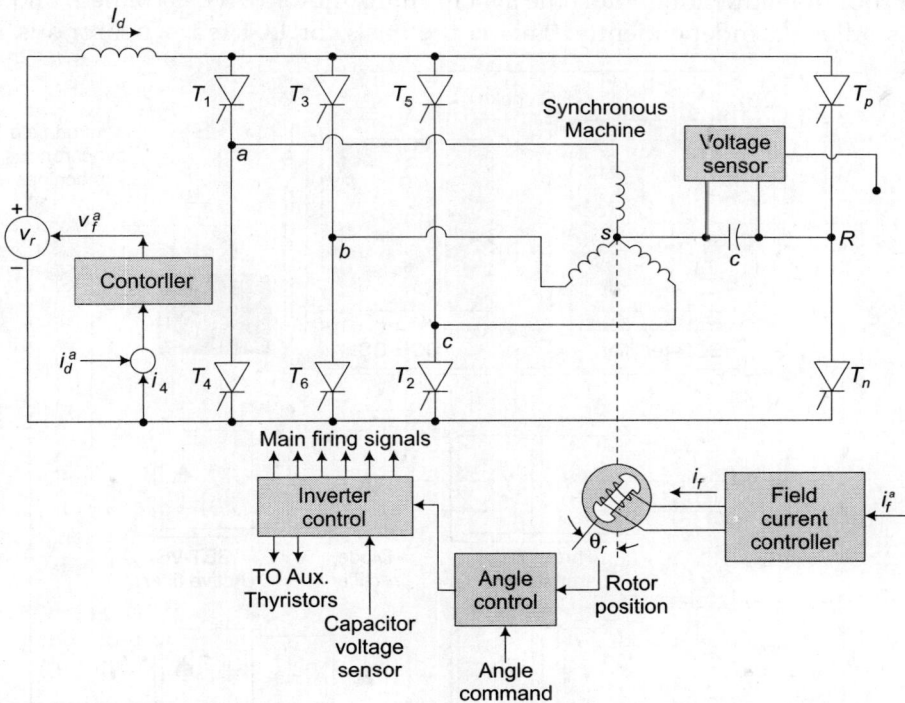

Fig. 9.8 Circuit configuration of LCI fed synchroncus motor.

The steady state performance of a line commutated inverter fed synchronous motor under variable frequencies has been determined. The triggering pulses for six SCRs of the three-phase inverter are generated by the control circuit in proper sequence by sensing terminal voltages of the synchronous motor, their similar characteristics of a DC series motor.

9.5.2   VSI Fed Synchronous Motor Drive

To reduce the speed of the rotating magnetic field of the stator to a low enough value that the rotor can easily accelerate and lock in with it during one half-cycle of the rotating magnetic field's rotation. This is done by reducing the frequency of the applied electric power. This method is usually followed in the case of inverter-fed synchronous motor operating under variable speed drive applications.

The main active power of the drive flows through line/load commutated SCR circuits switching at the fundamental frequency while the harmonics and reactive power are alone handled by the VSI. Since the reactive power requirement of high power motors is quite small. The kVA rating of the VSI is less as compared to the motor rating. An IGBT based VSI rated at the motor kVA has to be used with an output LC filters for the same application to obtain equivalent performance in terms of voltage and current. Since the switching frequency of such a high power IGBT will be limited, the design of the filters is quite difficult. The LCI + VSI converter configuration can be applied to other motors as well. From the induction motor, the synchronous motor drive is rotated, and the field current is adjusted independently. This is the basis for LCI fed synchronous machine

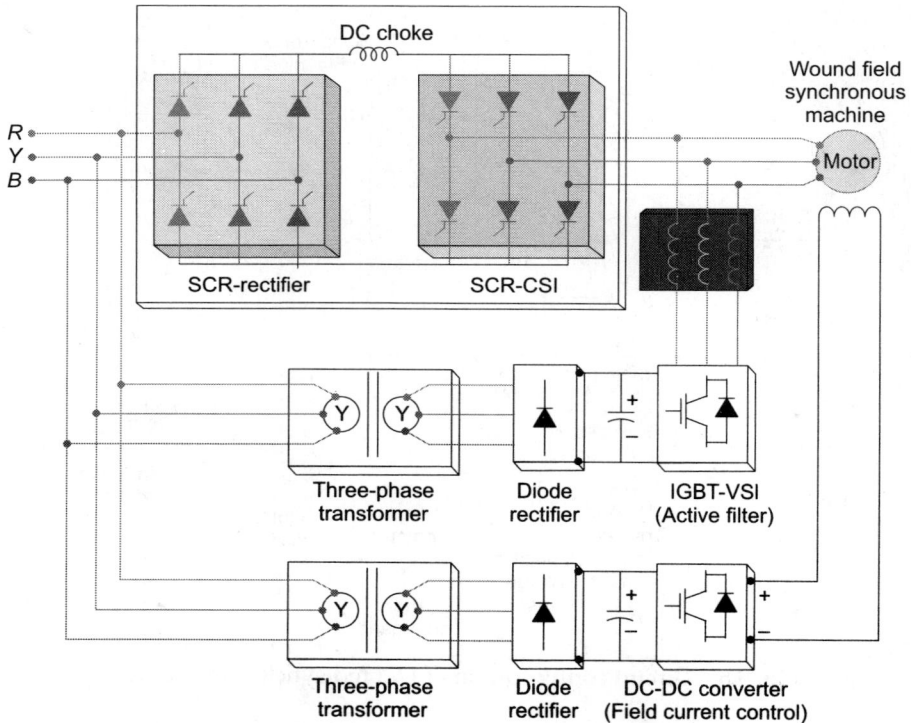

Fig. 9.9 VSI fed synchronous motor.

drives: the field current is adjusted to make the motor operate at leading displacement factor, thereby enabling load commutation of the CSI. However, the field circuit is heavily inductive and slow to respond. Therefore, use of a VSI is still advantageous, in terms of injecting the reactive power required under dynamic conditions, until the field circuit is able to take over. Moreover, the compensation of the harmonic currents of the CSI is still a great benefit. The currents and voltages of the motor are close to sinusoidal. The LCI fed synchronous machine also requires a special starting technique through DC link current pulsing. The motor can be started up smoothly by the VSI, even if the load torque is high. Thus, the addition of a VSI to the existing LCI drives as a retrofit can solve the problem of starting. The VSI can be disconnected from the motor once the speed reaches a minimum value. This would also limit the kVA rating of the VSI. The VSI can also be applied as a retrofit active filter to the existing LCI drives, in order to filter out the current harmonics and improve the motor current and voltage. The VSI can be applied and removed "on-the-fly" while the motor is being driven by the LCI. Thus, the LCI+VSI converter can be a universal solution for synchronous motor drives.

9.5.3   Cycloconverter Fed Synchronous Motor Drives

The cycloconverter-fed salient synchronous machine is very important in the applications of high power and low-speed drives, such as mining hoist and rolling mill. Full digital controllers and software are the most important core in the whole system. The calculation and communication capabilities of conventional controllers are not enough. It is applied in the vector control system of a cycloconverter-fed salient synchronous machine.

To increase the efficiency and quality of the grinding process, it is normal practice to use power converters to control the energy delivered to the synchronous motor.

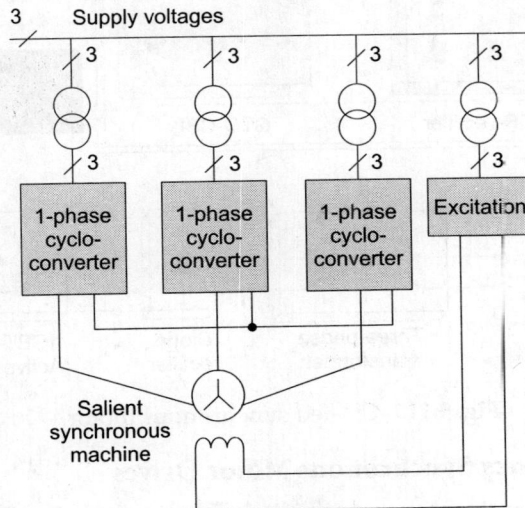

Fig. 9.10   Cycloconverter fed synchronous motor.

The cycloconverter has been the selected alternative, mainly due to its high efficiency and global performance. Because of the high power, these units have a big impact on the operation of the power system, and so the interaction with the power supply must be studied carefully. Blocking the gate pulse of the thyristors is an effective way to control drive. It works satisfactorily in an abnormal operating condition. The most appropriate way to provide the proper controlling in the connection of the cycloconverter is to block the gate pulses of the thyristors.

High power, wound-field synchronous motors can be operated at unity power factor when excited by phase-controlled, line-commutated, thyristor cycloconverter drive control for such drives can be both scalar and vector control, similar to that of the voltage-fed inverter drive.

9.5.4   CSI-Fed Synchronous Motor Drives

High power AC motor drives are the enabling technology in many industrial processes. The types of motor, converter and control method are important for selecting a drive for an application. The choices have been fairly stable for quite some time. Recent developments in power semiconductor devices have made it possible to propose newer configurations, which allow one to move closer to the ideal of delivering sinusoidal voltage as well as current to the motor.

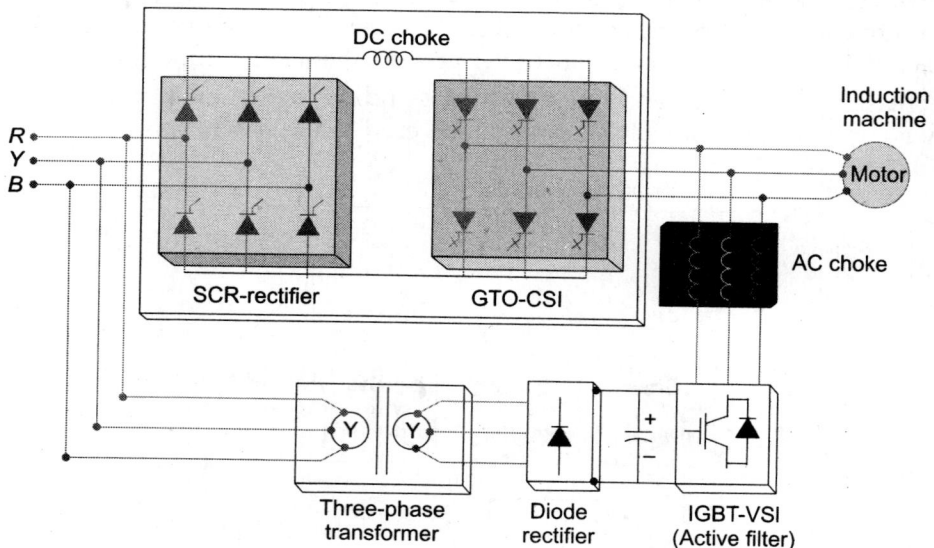

Fig. 9.11   CSI-fed synchronous motor.

9.5.5   Variable Frequency Synchronous Motor Drives

Synchronous motors operate at synchronism with the line frequency and maintain a constant speed regardless of load without sophisticated electronic control. The two most common types of synchronous motors are reluctance and a permanent magnet. The synchronous motor typically provides up to a maximum of 140% of rated torque. These designs start like an induction motor but quickly accelerate from approximately 90% sync speed to synchronous speed. When operated from an AC drive they require boost voltage to produce the required torque to synchronize quickly after power application.

9.5.6   Closed Loop Operation of Synchronous Motor Drives

- Measure the motor quantities (phase voltages and currents).
- Transform the quantities into a 2-phase system (α, β), using Clarke transformation.
- Calculate the rotor flux space vector magnitude and position angle.
- Transform stator currents into the d-q coordinate system using Park transformation.
- The stator current torque- (i_q^s) and flux- (i_d^s) producing components are separately controlled by the controllers.
- The output stator voltage space vector is calculated using the decoupling block.
- The stator voltage space vector is transformed back from the d-q coordinate system into the two-phase system and fixed with the stator by inverse Park transformation.
- Using sine wave modulation, the output 3-phase voltage is generated.

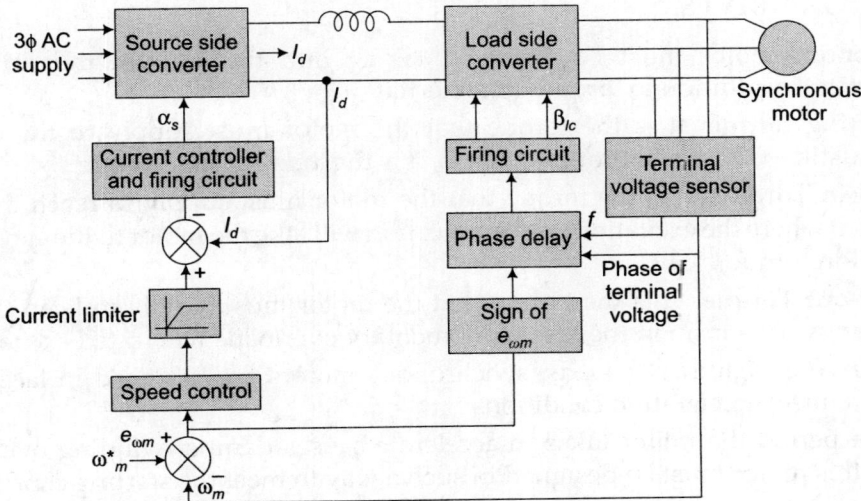

Fig. 9.12   Closed loop system of synchronous motor drives.

9.6   BRUSHLESS DC EXCITATION

Wound-field synchronous motors (WFSM) require DC current excitation in the rotor winding. This excitation is traditionally done through the use of slip rings and brushes. However, these have several disadvantages such as requiring maintenance, arcing.

A wound-rotor induction motor is mounted on the same shaft as the wound-field synchronous motor. This acts like a rotating transformer with the rotor as the primary and the stator as the secondary. The stator of the WFSM is fed by a supply frequency, and the rotor of the WFSM rotates at a speed set by the supply frequency. The slip voltage in the rotor winding of the WFSM is rectified to provide the current feed to the rotor windings of the synchronous motor.

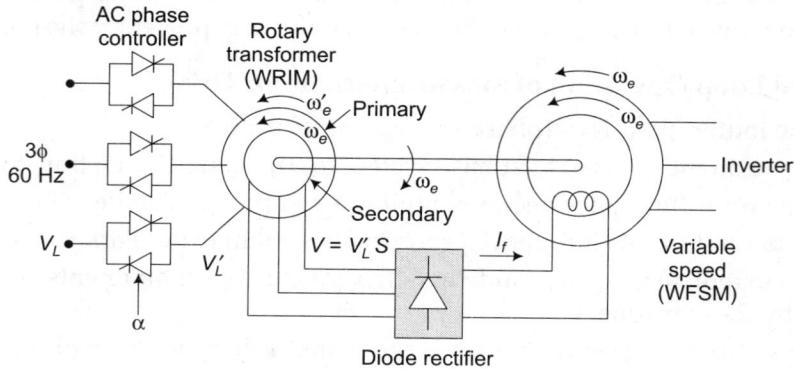

Fig. 9.13 Brushless DC excitation of WFSM.

9.7 SPEED-TORQUE CHARACTERISTICS OF SYNCHRONOUS MOTOR DRIVES

A synchronous motor must be designed taking into the account the driven load characteristics, in addition to torques and inertia.

(a) **Starting torque:** It is the torque that the motor must supply to the drive the standstill load resistant torque, that is, it is the load starting torque.

(b) **Pull-in Torque:** It is the torque that the motor must supply to reach the correct speed, where the excitation field application will take the motor to the synchronism (pull-in torque).

(c) **Pull-out Torque:** It is the torque that the motor must supply to keep the motor under synchronism in the case of momentary overloads with rated excitation.

When driving high inertia loads, synchronous motors are designed in larger frame sizes so as to meet acceleration conditions.

The time period the motor takes to accelerate causes damper winding overheating. Therefore, this motor must be designed in such a way to meet the starting conditions.

The correct load inertia definition, associated with motor and load torque analysis are quite important allowing this motor to meet starting and acceleration conditions.

The damper winding that operates as a squirrel cage of an induction motor is intended to guarantee synchronous motor starting and acceleration. This way, starting and pull-in torques vary with the square of the applied voltage, and the starting current is proportional to the applied voltage, exactly as on induction motors.

9.8 APPLICATION AND ADVANTAGES

Due to their special operating characteristics, synchronous motor applications usually result in economical and operational advantages to end users. Economic advantages of using synchronous motors are:

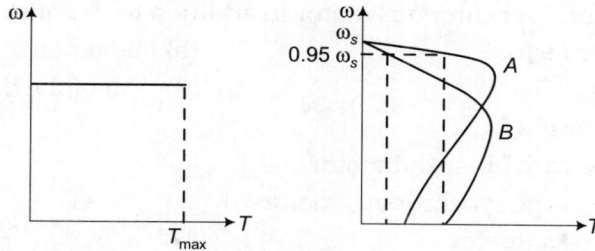

Fig. 9.14 Torque-speed characteristics of synchronous motor.

- High efficiency
- Power factor correction

In addition to that, there are other specific operational advantages of using synchronous motors as follows:

- Special starting characteristics
- Constant speed under load variation
- Reduced maintenance cost

Applications of synchronous motors require precise speed or position control:

1. Speed does not depend on the load over the operating range of the induction motor and DC motor.
2. In open loop system, speed and position may be accurately controlled.
3. Low-power applications of the synchronous machine where high precision is required.
4. A DC current is applied to both the stator and the rotor windings then they will hold their position.
5. For low-speed applications, efficiency is increased.

Multiple-Choice Questions

1. Synchronous motor can be controlled by power modulator:
   (a) Inverter   (b) Cycloconverter
   (c) AC voltage controller   (d) All of the above
2. Synchronous motor can be started by:
   (a) Damper winding   (b) Three point starter
   (c) Star-Delta starter   (d) (b) and (c)
3. It is the torque that the motor must supply to keep the motor under synchronism in the case of momentary overloads with rated excitation.
   (a) Electromagnetic torque   (b) Pull-up torque
   (c) Load torque   (d) Pull-out torque

4. The advantage of a synchronous motor in addition to its constant speed is:
    (a) High power factor
    (b) Better efficiency
    (c) Lower cost
    (d) All of the these

5. Reluctance motor is a...........
    (a) Low torque variable speed motor
    (b) Self-starting type synchronous motors
    (c) Variable torque motor
    (d) Low noise, slow speed motor

6. A VSI is normally employed when
    (a) Source inductance is large, and the load inductance is small
    (b) Source inductance is small, and the load inductance is large
    (c) Both source inductance and the load inductance are small
    (d) Both source inductance and the load inductance are large

7. A synchronous motor is found to be more economical when the load is above
    (a) 100 kW
    (b) 1 kW
    (c) 10 kW
    (d) 20 kW

8. A line commutated phase controlled is operating at its inverter limit. There can be a commutation failure if
    (a) The frequency increases
    (b) The voltage increases
    (c) Both frequency and voltage change but its ratio will be constant
    (d) The frequency decreases

9. The motor has the least range of speed control.
    (a) Slip ring induction motor
    (b) Synchronous motor
    (c) DC shunt motor
    (d) Reluctance motor

Answers

1. (d)	2. (a)	3. (d)	4. (a)	5. (b)
6. (b)	7. (a)	8. (d)	9. (b)	

Exercise

1. Write a short note on self-control mode of wound-rotor synchronous motor drives.
2. Draw the block diagram of closed loop synchronous motor drives.
3. Explain the starting methods of the synchronous motor.
4. Explain LCI fed synchronous motor drives.

5. Explain the methods of closed loop simulation of synchronous motor drives.

6. What are the advantages of the synchronous motor over induction motor?

7. What are the applications of synchronous motor drives?

8. Explain the difference between separate and self-control of synchronous motor.

9. Explain the operation of VSI fed self-controlled synchronous motor drive with neat circuit and waveforms and also draw their speed torque characteristics.

10. Explain the operation of CSI fed self-controlled synchronous motor drive with neat circuit and waveforms and also draw their speed torque characteristics.

11. Explain the operation of cycloconverter fed self-controlled synchronous motor drive with neat circuit and waveforms and also draw their speed torque characteristics.

12. Explain the closed loop control of synchronous motor drives with neat block diagrams.

13. Describe the advantages, disadvantages and applications of self-controlled synchronous motor fed by VSI, CSI, and cycloconverter.

14. Describe the operation of LCI fed synchronous motor drive.

15. Describe the working of CSI fed synchronous motor.

10

Industrial Applications of Electric Drives

10.1 INTRODUCTION

Electrical drives are used in industries to maintain the quality of production as per the requirement. The energy consumption should be minimum. Air clutch and synchronous motor can be used for individual drive to achieve higher efficiency. The first Gearless Mill Drive (GMD) with a power of 6.5 MW was installed in 1969 in a cement plant, Le Havre in France. Nowadays, the traditionally used direct current (DC) drive has been replaced by induction motor drives. A system incorporating AC motor has low investment cost higher efficiency. Selection of drives and motors for powering the kiln, rolling mills, and cement mills according to applications is an evolving process.

Large capacity motors and power electronic devices are used in processing of steel in an integrated steel plant. Raw material preparation, primary reduction, refining, casting, hot and cold rolling, surface coating are the processes of manufacturing steel products from iron ore. Gas and electric power in the form of energy is required in each of these processes. In finishing mills, generally motors used require speed regulation of high accuracy. In most of the steel plant applications, variable speed was obtained through DC motors driven four-quadrant converters.

Nowadays, the DC motors and drives are replaced by AC motors and drives. Large AC power is also used for induction heating in ladle furnace during secondary refining of steel, high-frequency heating during welding process in a pipe plant.

10.2 CLASSIFICATION OF INDUSTRIAL DRIVES

- **Rolling Mills:**
  (1) Reversing hot rolling mills
  (2) Continuous hot rolling mills
  (3) Reversing cold rolling mills
  (4) Continuous cold rolling mills
- **Kiln drives:**
Pyroprocessing devices used in a rotary kiln for raising the temperature of materials to a higher level.
- The limestone mined from the quarries is crushed and transported to the plant by dumpers wagons, trucks or ropeways depending on the area and distance involved.
- The quarry is within 1-2 km and in the line of supply of the limestone. The crushed limestone together with the required proportion of corrective additives like clay bauxite, iron, etc., is ground in grinding mills.

- The fine dry powder coming out is homogenized in silos by the passage of air from the bottom and through the medium. It is then fed into the kiln, which is the heart of the cement plant, for producing cement clinker at high temperatures.
- The kiln receives finally ground and exactly composed dry feed as mentioned above; the cement plant is called a dry process one. In the wet process, the raw materials are ground with water to produce semi-liquid mixture before entering the kiln feed tank.
- The dry process is preferred to wet process because less fuel is required by such kilns.
- Wet process is sometimes necessitated , since certain materials contain so much water that using wet process is better than trying to dry the raw materials.
- The clinker coming out of the kiln is air cooled in special types of coolers and then transported to the storage. After aging in storage for at least three days the clinker, mixed with the right amount of gypsum is fed to the cement grinding mills and ground to required fineness. Cement is stored in silos, drawn for packing in gunny bags and dispatched by wagons or trucks to dealers.

Classification of driving motor used in cement industry are as follows:

(1) Raw mill and cement mill drives
(2) Crusher drives
(3) Waste gas fan drives
(4) Compressor drives

10.3   CEMENT MILL DRIVES

AC drives are generally used to maintain the speed or torque of AC motors. Induction motor drives replace DC motors and slip-ring motors along with their control systems. Induction motor drives also replace the need for starters, cascade drives, hydraulic speed control, mechanical gears, fan inlet vane control, fan damper control and many other techniques of regulating the speed of electric motors used throughout the cement making process. Presently, AC drives are used in a wide range of applications, some of which involve much more than just rotating the motor.

The advantages include:
- Energy consumption is very low.
- Minimum pollution.
- Less investment in electrical network compensation devices, such as filters.
- Harmonic distortion can be minimized to the electricity supply network.
- Higher process reliability and quality.
- Better efficiency.
- Increased productivity.

Applications:
- Variable frequency drives (VFDs) are used to start large motors and continuously adjust the speed as required by the process.

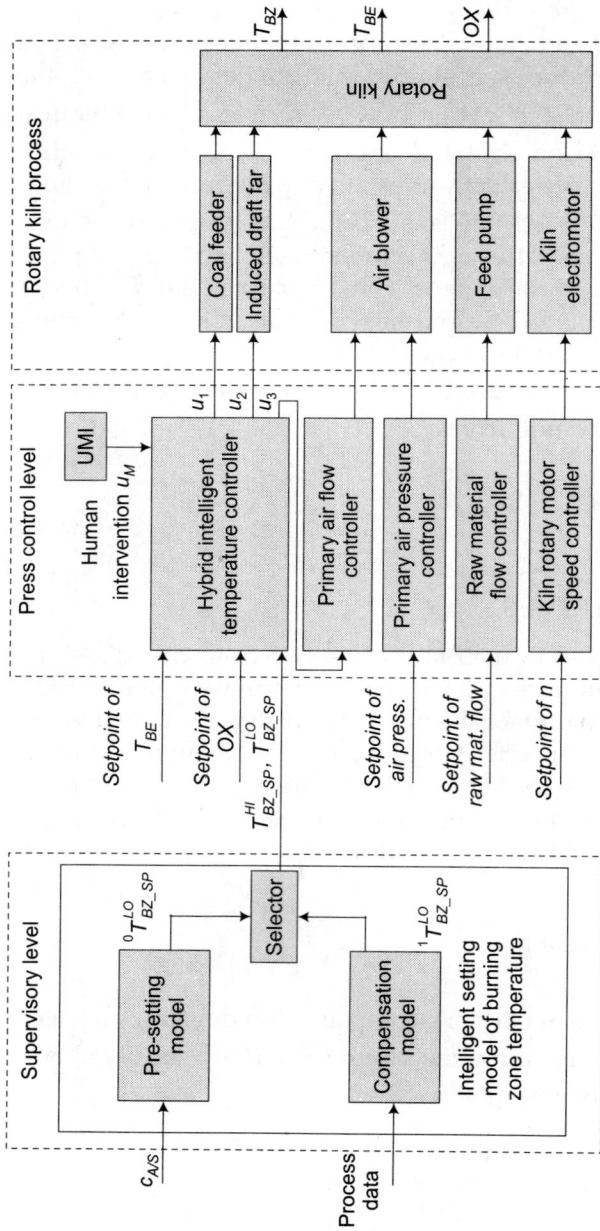

Fig. 10.1 Processing of kiln drives.

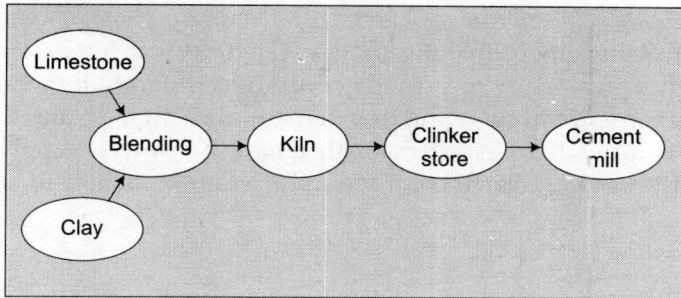

Fig. 10.2   Processing of cement mill.

• Induction and synchronous motors driving excavators, crushers, conveyors, mills, kilns, and fans use VFDs to provide high power, speed control, and low-loss flow control with significantly associated energy savings.
• By controlling the speed and torque of the crusher motor by means of a low voltage, AC drives the lifetime of the equipment can be extended.

Conveyors:

In cement making process, many conveyors are used, with the risk of belts stretching, slipping or breaking. A low voltage AC drive protects the belts and other mechanical equipment by offering smooth and accurate control of the motor.

Feeders:

The quality of cement depends on the accurate and estimated quantity of raw materials and additives. It is also necessary to avoid stretching, slipping and breaking of feeder belts. Controlling the speed and torque of feeder motors with low voltage AC drives ensures precise material dosage which is under continuous control. AC drives enable exact information on the estimated amounts of materials.

Grinders:

Grinding raw material and clinker causes huge wear to the grinding mill. Starting the grinder, direct-on-line, stresses the grinder and the gearbox, increasing the risk of gearbox failure as well as shortening the lifetime of the mechanical equipment. The use of low voltage AC drives helps to control the grinder speed to match the material flow, and minimize the wear of the grinder. Also, the mechanical stresses during starting are eliminated. Consumption of energy is high in grinding. The efficiency of grinding mills is low. AC drives provide efficient use of energy and improve the overall efficiency of the grinding process.

Separators:

In the cement making process, separators are crucial and have a major impact on the quality of cement as well as energy consumption. The overall process characteristics include gas flows, size of particles and require precise and rapid speed control of the separator motors. In order to gain rapid speed reductions or to stop the separators the high inertia of the separators requires braking. Low voltage AC drives provide fast and accurate separator speed control.

Kilns:

Kilns play a key role in cement making process. Therefore, it requires reliable and high-performance motors and drives to provide continuous operation in varying conditions. This is the stage where capital costs and fuel demands are highest, and the process control is crucial. There is a need to restart the kiln when the power supply fails. Restarting of kiln requires a very high starting torque, which can be achieved with direct torque control (DTC).

Fans:

There is always a need for precise control of the gas flow. Fans are used for both cooling the air and the exhaust gases produced by the cement making process itself.

Smooth control of the gas flow has an important role in maintaining a consistent cement quality. Fans which are efficiently controlled prove are the biggest source of energy saving. Controlling the fan speed with AC drives is the most energy-efficient control method, providing significant energy savings compared to any other control method.

Both AC and DC drives are employed for the different operations in a crane. The preferred drives on consideration of economy end utility are indicated below:

Operation	Type of drive
(i) Hoisting and lowering	AC slip ring motor. Ward Leonard. Controlled DC shunts motor and DC compound motor.
(ii) Crane travel	AC slip ring motor.
(iii) Trolley travel	AC slip ring motor.
(iv) Slew and swing action	AC slip ring motor or DC shunt motor.
(v) Boom hoist	AC slip ring motor.

10.4   ROLLING MILL DRIVES

The following types of drives are used for rolling mills:
   (i)   DC motors
   (ii) AC slip ring motors with speed control.

The DC motors, because of their inherent characteristics, are best suited for the rolling mills. Speed control is affected either through Ward Leonard system or by grid controlled mercury arc rectifiers. AC slip ring motors are suitable for roughing, and re-rolling mills where very precise speed control is not required. Their efficiency is low because of the power wasted in the rotor resistance. There is also abrupt rise in motor speed when the material leaves the rolling stands.

10.5   KILN DRIVES

Call for a starting torque of about 250% in addition to the speed control feature. The commonly used drives are:
   (i)   Slip ring induction motor.
   (ii) Three-phase shunt wound commutator motor.

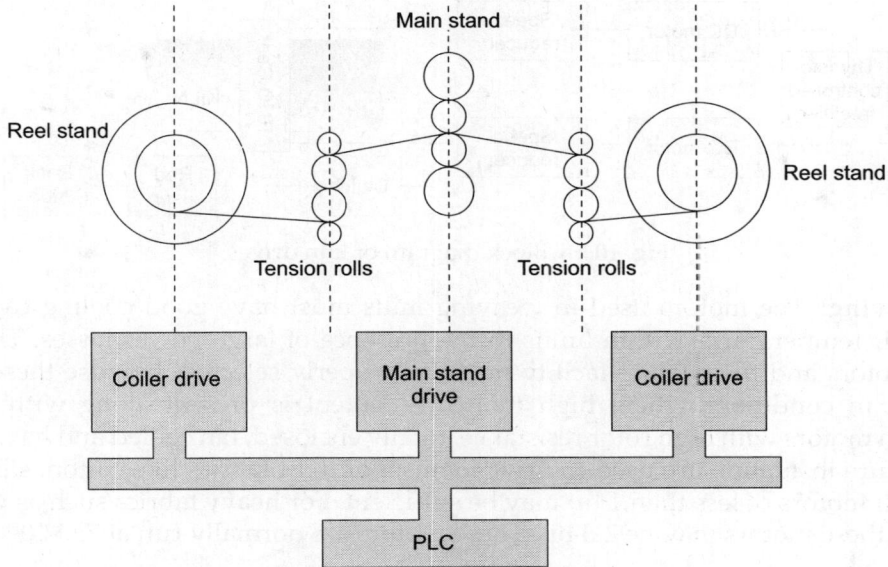

Fig. 10.3   Rolling mill drives.

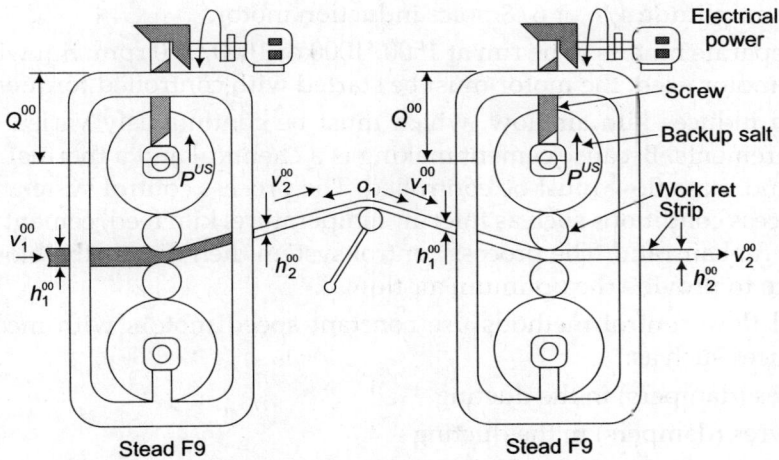

Fig. 10.4   Rolling mill process.

  (iii) Cascade controlled AC motor.

  (iv) Ward Leonard controlled DC motor.

  (v) DC motor with transformer step switch control.

10.6   TEXTILE INDUSTRY

The textile industry requires special types of drives for:

  (i) weaving, and

  (ii) spinning.

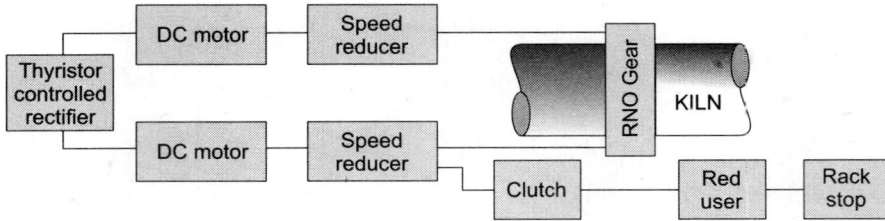

Fig. 10.5   Block diagram of kiln drives.

(i)   Weaving: The motors used in weaving mills must have good cooling capacity to keep their temperatures within limits in the presence of large power losses. The rating of the motors and the cooling facility must be properly selected, because these motors are used in conditions where high moisture content is present along with lot cage induction motors with high rotor resistance, totally enclosed, fan cooled and having high-temperature insulation are used to drive looms. For light fabrics like cotton, silk, nylon, etc., small motors of less than 1 hp may be sufficient. For heavy fabrics such as wool, the rating of these motors may be 2-3 hp. These motors are normally run at 750-1000 rpm.

(ii)   Spinning: Spinning mills use one of the following three types of drives:

   (i)   A 4-pole or 6-pole squirrel-cage induction motor.
   (ii)   A pole amplitude 4/6 or 6/8 poles induction motor.
   (iii)   Two separate motors to be run at 1500/1000 or 1000/750 rpm. But whichever may be the motor used, the motor must be started with controlled torque.

The ID fan induces kiln air flow, which must be continuously varied to match the process requirements. Because cement making is a chemical and a thermal process, both air volume, and mass flow must be controlled. The process control system continuously monitors process conditions such as inlet air temperature, kiln feed, cement composition, and required fuel air ratio. The process control system then directs the blower and flow control system to provide the optimum air flow.

Traditional flow control methods use constant speed motors with mechanical flow reducing devices such as:

• Inlet louvres (dampers) in the ducting
• Outlet louvres (dampers) in the ducting
• Flow guide vanes in the fan casing
• Variable slip clutches in the fan drive shaft
• These mechanical solutions have significant disadvantages:
• High energy consumption at reduced flow rates
• Mechanical wear and required maintenance
• Process interruptions due to mechanical problems
• Limitations on motor starting duty

10.7   PAPER INDUSTRY

In a paper industry, the drives are required for

   (i) Pulp making, and

  (ii) Paper making.

In the paper making process, the logs of wood are either ground in mechanical grinders or else they are chemically treated with alkalis and simultaneously beaten up to turn them into a soft pulp. In the mechanical method of pulp making, the electrical power requirement is very high because the wood is hard. Since the mechanical grinders operate at a constant speed of about 200-300 rpm. The motors can be started on no load. Thus, synchronous motors are used for these drives. These motors normally run at 1000-1500 rpm and gears are used to reduce the speed to 200-300 rpm.

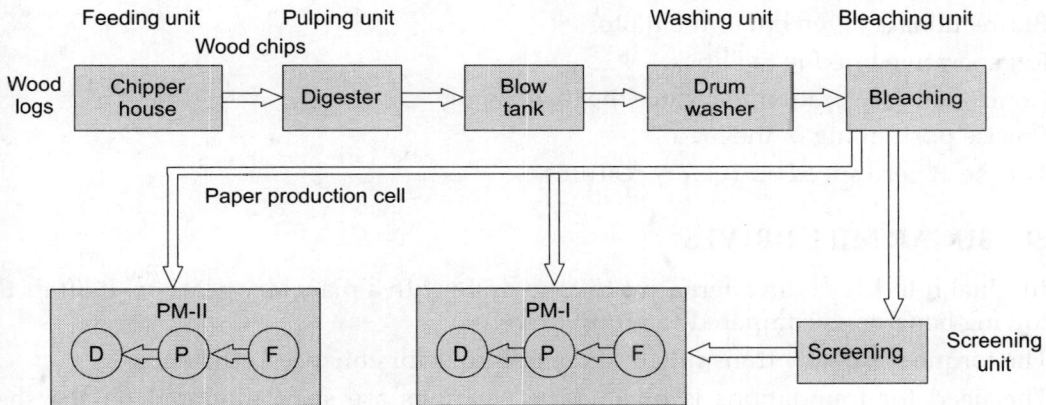

Fig. 10.6   Block diagram of paper production.

In the chemical method of pulp making, the logs of wood are continuously beaten by beaters at the time of treatment with alkali. The power requirement of the beater motors is less than those of grinder motors, but these motors require high starting torque. Therefore, slip ring induction motors with gears are used to drive these beaters at about 150-200 rpm.

Variable frequency drives for the paper industry.

Pumping System

- When in use, the variable frequency drives are always part of a pumping system. A pumping system is usually a network of pipes, tanks, valves and other system parts.

- The receiver is usually at a higher geographic level than the supply of the system. These parts can also be on the same level, as in the case of a closed circuit heat transfer system.

- Pumping systems mostly require a variation of flow rate. Examples include the daily cycle in the consumption of drinking water, the varying process demand for a liquid or seasonal heating demand.
- The variation required may be in the pump head, such as for cyclical changes in process pressure, or pumping to tanks with a variable liquid level. In spite of the variations, the pump capacity is selected according to the maximum flow and head or even to the future needs, perhaps with a certain safety margin.
- The average pumping capacity may be only a fraction of the maximum capacity, and this will require some kind of control.

10.8   CRANE DRIVES

The basic requirements of crane drives are:
- Bidirectional movement of the motors
- Regenerative braking facilities
- Controlled acceleration to reduce the load swing
- Precise positioning of the load
- Torque at zero speed and safety features

10.9   SUGAR MILL DRIVES

Individual roll drive is an alternative to cane drive with a planetary gearbox. It offers the following benefits as compared to group drive.
- The torque is directly transmitted to the mill rolls for enhanced efficiency.
- The need for foundations is obviated as gearings are shaft mounted on the shaft journals.
- There is no effect of top roll displacements on the operational function and wear of the drive as the drive follows the movement of the roll without any impediment from other factors.

The majority of drives in a sugar factory are of the constant speed type for which the three-phase induction motor is ideally suited. The squirrel cage induction motor is used for most normal applications while the slip ring induction motor comes into its own where heavy starting torque and/or prolonged run-up time is required. These motors are of a very robust design and require less maintenance. The control gear is simple and compact, thus ensuring reliability in an average installation.

The most important application of this type of drive is that of driving cane milling units, where the basic requirements are constant torque over the speed range. It is of advantage to have a reserve of torque should it be required. It is rather surprising that an industry as large as the sugar industry should resort to the steam turbine as a means of obtaining a variable speed drive, whereas most other major industries rely solely on electric drives.

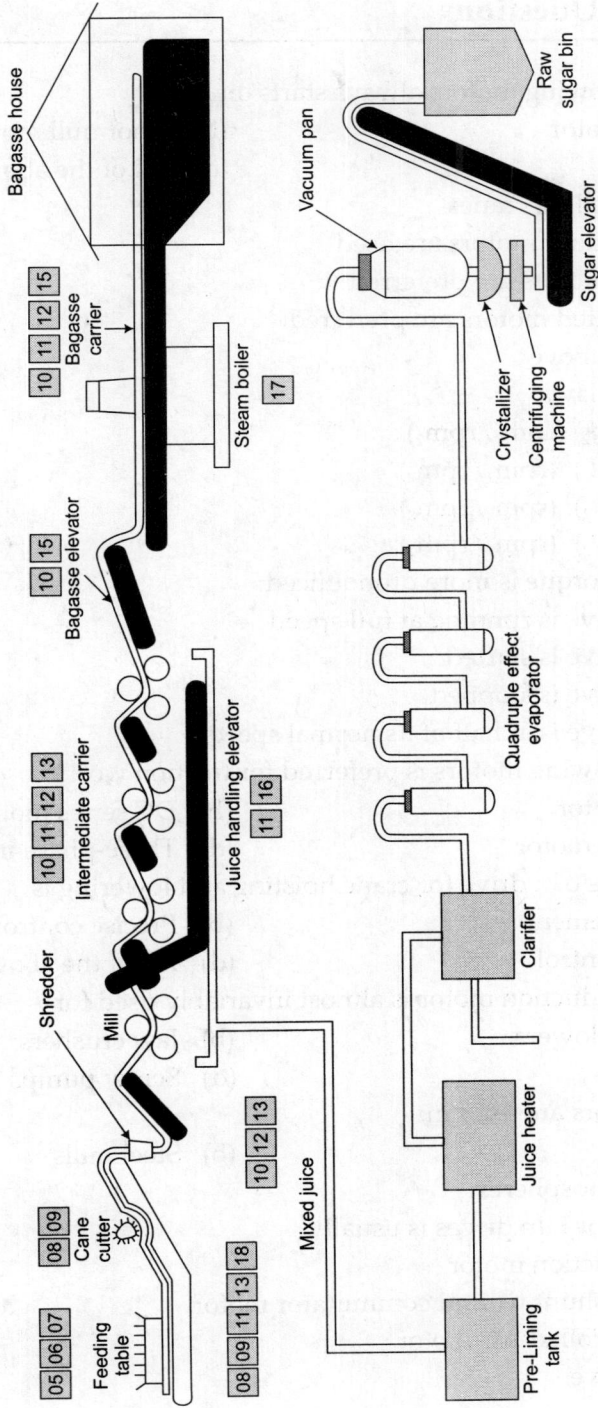

Fig. 10.7 Sugar mill.

Bagasse house

10 11 12 15
Bagasse carrier

Steam boiler
17

10 15
Bagasse elevator

10 11 12 13
Intermediate carrier

Shredder

Mill

11 16
Juice handling elevator

08 09
Cane cutter

05 06 07
Feeding table

08 09 11 13 18

10 12 13
Mixed juice

Vacuum pan

Raw sugar bin

Sugar elevator

Crystallizer
Centrifuging machine

Quadruple effect evaporator

Clarifier

Juice heater

Pre-Liming tank

Multiple-Choice Questions

1. Which of the following motors always starts on load
   - (a) Conveyor motor
   - (b) Floor mill motor
   - (c) Fan motor
   - (d) All of the above
2. In overhead travelling cranes
   - (a) Continuous duty motors are used
   - (b) Slow speed motors are preferred
   - (c) Short time rated motors are preferred
   - (d) None of the above
3. According to fan laws
   - (a) $V_1/V_2 = (d_1/d_2) (rpm_1/rpm_2)$
   - (b) $V_1/V_2 = (d_1/d_2)^2 (rpm_1/rpm_2)$
   - (c) $V_1/V_2 = (d_1/d_2)^3 (rpm_1/rpm_2)$
   - (d) $V_1/V_2 = (d_1/d_2)^3 (rpm_1/rpm_2)$
4. Effect of friction torque is more pronounced
   - (a) When the drive is running at full speed
   - (b) When the drive is started
   - (c) When the drive is stopped
   - (d) When the drive is at half of its normal speed
5. Which of the following motors is preferred for traction work?
   - (a) Universal motor
   - (b) DC series motor
   - (c) Synchronous motor
   - (d) Three-phase induction motor
6. The characteristics of a drive for crane hoisting and lowering is
   - (a) Smooth movement
   - (b) Precise control
   - (c) Fast speed control
   - (d) All of the above
7. Belted slip ring induction motor is almost invariably used for
   - (a) Centrifugal blowers
   - (b) Jaw crushers
   - (c) Water pumps
   - (d) Screw pumps
8. Flame-proof motors are used in:
   - (a) Paper mills
   - (b) Steel mills
   - (c) Explosive atmospheres
9. Motor preferred for kiln drives is usually.
   - (a) Slip ring induction motor
   - (b) Three-phase shunt wound commutator motor
   - (c) Cascade controlled AC motor
   - (d) All of the above

10. Which of the following motor always starts on load
    - (a) Conveyor motor
    - (b) Floor mill motor
    - (c) Fan motor
    - (d) All of the above

Answers

1. (d)	2. (c)	3. (c)	4. (b)	5. (b)
6. (d)	7. (b)	8. (c)	9. (d)	10. (d)

Exercise

1. Explain the industrial applications of electric drives.
2. Explain DC motor drive used in rolling mill.
3. Give the classification of industrial drives in brief.
4. What are the basic requirements of crane drives?
5. Explain the different processes used in cement making.
6. What are the benefits and applications of VFDs in industrial drives?
7. Explain weaving and spinning processes used in textile industry.
8. Explain sugar making process in a sugar mill.
9. Explain the chemical method of pulp making with block diagram in the paper industry.
10. Explain the applications of ASD in rolling mills.
11. Write a short note on ASD used in textile mills.
12. Explain application of drives in sugar mills.
13. Explain how ASD can be used in machine tool applications.
14. State drive requirements of reversing cold rolling mills.
15. Suggest suitable type of drives for following applications and justify the answer:
    - (i) Pumps and blowers
    - (ii) Compressors
    - (iii) Kiln drives in cement mills
16. Describe the role of drives in textile mills.
17. State the requirements of motors used for machine tools.

A

Appendix

MATLAB Code for Speed Control of 3-Phase Induction Motor Using Variable Rotor Resistance

```
Vll=input('Enter the Supply Voltage (line to line) RMS value:');
P=input('Enter the number of poles: ');
Rs=input('Stator Resistance: ');
Rr1=input('Enter the first Rotor Resistance: ');
Rr2=input('Enter the second Rotor Resistance: ');
Rr3=input('Enter the third Rotor Resistance: ');
Rr4=input('Enter the fourth Rotor Resistance: ');
Rr5=input('Enter the fifth Rotor Resistance: ');
Xs=input('Stator Leakage Reactance @ 60 Hz frequency: ');
Xr=input('Rotor Leakage Reactance @ 60 Hz frequency: ');
Ls=Xs/(2*pi*60);
Lr=Xr/(2*pi*60);

Wsync1=4*pi*60/P;
Tmf2=zeros(Wsync1*500+1,1);
Tmf3=zeros(Wsync1*500+1,1);
Tmf4=zeros(Wsync1*500+1,1);
Tmf5=zeros(Wsync1*500+1,1);
Tmf1=zeros(Wsync1*500+1,1);
m=1;
 for Wrotor1=0:0.002:Wsync1
 Tmf1(m)=(3*(((Vll^2)*Rr1/((Wsync1-Wrotor1)/Wsync1))/((Rs+Rr1/
((Wsync1- Wrotor1)/Wsync1))
^2+(2*pi*60*Ls+2*pi*60*Lr)^2))/Wsync1); %star connected m=m+1;
end
m=1;
 for Wrotor1=0:0.002:Wsync1
    Tmf2(m)=(3*(((Vll^2)*Rr2/((Wsync1-Wrotor1)/Wsync1))/((Rs+Rr2/
((Wsync1- Wrotor1)/Wsync1))
```

```
^2+(2*pi*60*Ls+2*pi*60*Lr)^2))/Wsync1);
m=m+1;
end
m=1;
  for Wrotor1=0:0.002:Wsync1
Tmf3(m)=(3*(((Vl1^2)*Rr3/((Wsync1-Wrotor1)/Wsync1))/((Rs+Rr3/
((Wsync1- Wrotor1)/Wsync1))
^2+(2*pi*60*Ls+2*pi*60*Lr)^2))/Wsync1);
m=m+1;
end
 m=1;
  for Wrotor1=0:0.002:Wsync1
    Tmf4(m)=(3*(((Vl1^2)*Rr4/((Wsync1-Wrotor1)/Wsync1))/((Rs+Rr4/
((Wsync1- Wrotor1)/Wsync1))
^2+(2*pi*60*Ls+2*pi*60*Lr)^2))/Wsync1);
m=m+1;
end
m=1;
  for Wrotor1=0:0.002:Wsync1
    Tmf5(m)=(3*(((Vl1^2)*Rr5/((Wsync1-Wrotor1)/Wsync1))/((Rs+Rr5/
((Wsync1- Wrotor1)/Wsync1))
^2+(2*pi*60*Ls+2*pi*60*Lr)^2))/Wsync1);
m=m+1;
end
plot(Tmf1);
hold on;
plot(Tmf2);
plot(Tmf3);
plot(Tmf4);
plot(Tmf5);
hold off;
ylabel('Torque(N-m)');
xlabel('Rotor Speed(Rad/s)');
end
```

B

Appendix

MATLAB Code for Speed Control of 3-Phase Induction Motor Using Variable Stator Voltage

```
Vl1=input('Enter the first Supply Voltage(line to line) RMS value: ');
Vl2=input('Enter the second Supply Voltage(line to line) RMS value: ');
Vl3=input('Enter the third Supply Voltage(line to line) RMS value: ');
Vl4=input('Enter the fourth Supply Voltage(line to line) RMS value: ');
Vl5=input('Enter the fifth Supply Voltage(line to line) RMS value: ');
P=input('Enter the number of poles: ');
Rs=input('Stator Resistance: ');
Rr=input('Rotor Resistance: ');
Xs=input('Stator Leakage Reactance @ 60 Hz frequency: ');
Xr=input('Rotor Leakage Reactance @ 60 Hz frequency: ');

Ls=Xs/(2*pi*60);
Lr=Xr/(2*pi*60);

Wsync1=4*pi*60/P;
Tmf2=zeros(Wsync1*500+1,1);
Tmf3=zeros(Wsync1*500+1,1);
Tmf4=zeros(Wsync1*500+1,1);
Tmf5=zeros(Wsync1*500+1,1);
Tmf1=zeros(Wsync1*500+1,1);
m=1;
 for Wrotor1=0:0.002:Wsync1
   Tmf1(m)=(3*(((Vl1^2)*Rr/((Wsync1-Wrotor1)/Wsync1))/((Rs+Rr/((Wsync1-
Wrotor1)/Wsync1))
^2+(2*pi*60*Ls+2*pi*60*Lr)^2))/Wsync1); %star connected m=m+1;
end
m=1;
 for Wrotor1=0:0.002:Wsync1
   Tmf2(m)=(3*(((Vl2^2)*Rr/((Wsync1-Wrotor1)/Wsync1))/((Rs+Rr/((Wsync1-
Wrotor1)/Wsync1))
^2+(2*pi*60*Ls+2*pi*60*Lr)^2))/Wsync1);
m=m+1;
end
m=1;
```

```
for Wrotor1=0:0.002:Wsync1
    Tmf3(m)=(3*(((Vl3^2)*Rr/((Wsync1-Wrotor1)/Wsync1))/((Rs+Rr/((Wsync1-Wrotor1)/Wsync1))
^2+(2*pi*60*Ls+2*pi*60*Lr)^2))/Wsync1);
m=m+1;
end
m=1;
 for Wrotor1=0:0.002:Wsync1
    Tmf4(m)=(3*(((Vl4^2)*Rr/((Wsync1-Wrotor1)/Wsync1))/((Rs+Rr/((Wsync1-Wrotor1)/Wsync1))
^2+(2*pi*60*Ls+2*pi*60*Lr)^2))/Wsync1);
m=m+1;
end
m=1;
 for Wrotor1=0:0.002:Wsync1
    Tmf5(m)=(3*(((Vl5^2)*Rr/((Wsync1-Wrotor1)/Wsync1))/((Rs+Rr/((Wsync1-Wrotor1)/Wsync1))
^2+(2*pi*60*Ls+2*pi*60*Lr)^2))/Wsync1);
m=m+1;
end
plot(Tmf1);
hold on;
plot(Tmf2);
plot(Tmf3);
plot(Tmf4);
plot(Tmf5);
hold off;
ylabel('Torque(N-m)');
xlabel('Rotor Speed(Rad/s)');
end
```

C

Appendix

MATLAB Code for Speed Control of 3-Phase Induction Motor Using Constant V/F Control

```
Vll=input('Supply Voltage (line to line) RMS value @ 60 Hz: ');
f2=input('Enter the second frequency: ');
f3=input('Enter the third frequency: ');
f4=input('Enter the fourth frequency: ');
f5=input('Enter the fifth frequency: ');
P=input('Enter the number of poles: ');
Rs=input('Stator Resistance: ');
Rr=input('Rotor Resistance: ');
Xs=input('Stator Leakage Reactance @ 60 Hz frequency: ');
Xr=input('Rotor Leakage Reactance @ 60 Hz frequency: ');
Ls=Xs/(2*pi*60);
Lr=Xr/(2*pi*60);
Vlnf1=Vll/(3^0.5);
Vlnf2=Vlnf1*f2/60;
Vlnf3=Vlnf1*f3/60;
Vlnf4=Vlnf1*f4/60;
Vlnf5=Vlnf1*f5/60;
Wsync1=4*pi*60/P;
Wsync2=4*pi*f2/P;
Wsync3=4*pi*f3/P;
Wsync4=4*pi*f4/P;
Wsync5=4*pi*f5/P;
Tmf2=zeros(Wsync2*500+1,1);
Tmf3=zeros(Wsync3*500+1,1);
Tmf4=zeros(Wsync4*500+1,1);
Tmf5=zeros(Wsync5*500+1,1);
Tmf1=zeros(Wsync1*500+1,1);
m=1;
  for Wrotor1=0:0.002:Wsync1
```

```
    Tmf1(m)=(3*(((Vlnf1^2)*Rr/((Wsync1-Wrotor1)/Wsync1))/((Rs+Rr/
((Wsync1-Wrotor1)/Wsync1))^2+(2*pi*60*Ls+2*pi*60*Lr)^2))/Wsync1);
%star connected
 m=m+1;
end
m=1;
for Wrotor2=0:0.002:Wsync2
Tmf2(m)=(3*(((Vlnf2^2)*Rr/((Wsync2-Wrotor2)/Wsync2))/((Rs+Rr/
((Wsync2-Wrotor2)/Wsync2))^2+(2*pi*f2*Ls+2*pi*f2*Lr)^2))/
Wsync2);
 m=m+1;
end
m=1;
for Wrotor3=0:0.002:Wsync3
    Tmf3(m)=(3*(((Vlnf3^2)*Rr/((Wsync3-Wrotor3)/Wsync3))/((Rs+Rr/
((Wsync3-Wrotor3)/Wsync3))^2+(2*pi*f3*Ls+2*pi*f3*Lr)^2))/Wsync3);
 m=m+1;
 end
 m=1;
 for Wrotor4=0:0.002:Wsync4
    Tmf4(m)=(3*(((Vlnf4^2)*Rr/((Wsync4-Wrotor4)/Wsync4))/((Rs+Rr/
((Wsync4-Wrotor4)/Wsync4))^2+(2*pi*f4*Ls+2*pi*f4*Lr)^2))/Wsync4);
 m=m+1;
 end
 m=1;
 for Wrotor5=0:0.002:Wsync5
    Tmf5(m)=(3*(((Vlnf5^2)*Rr/((Wsync5-Wrotor5)/Wsync5))/((Rs+Rr/
((Wsync5-Wrotor5)/Wsync5))^2+(2*pi*f5*Ls+2*pi*f5*Lr)^2))/Wsync5);
 m=m+1;
 end
 plot(Tmf1);
 hold on;
 plot(Tmf2);
 plot(Tmf3);
 plot(Tmf4);
 plot(Tmf5);
 hold off;
 ylabel('Torque(N-m)');
 xlabel('Rotor Speed(Rad/s) * 100');
end
```

D

Appendix

MATLAB Code to Observe the Variations in Q-Axis and D-Axis Stator Currents with Change in Stator Voltage for A 3-Phase Induction Motor

```
Vll=input('Supply Voltage (line to line) RMS value @ 60 Hz: ');
f2=input('Enter the second frequency: ');
f3=input('Enter the third frequency: ');
f4=input('Enter the fourth frequency: ');
f5=input('Enter the fifth frequency: ');
P=input('Enter the number of poles: ');
Rs=input('Stator Resistance: ');
Rr=input('Rotor Resistance: ');
Xs=input('Stator Leakage Reactance @ 60 Hz frequency: ');
Xr=input('Rotor Leakage Reactance @ 60 Hz frequency: ');
%Motor detalis (Rated)
 P = 50*746;
 V = 460;
 Va = 460/sqrt(3) ;
 fe = 60;
 We = 2*pi*fe;
 Lambda = sqrt(2)*Va/We;
 I= P/(sqrt(3)*V) ;
 Ipeakbase = sqrt(2)*I;
 poles=4 ;
%Paramteres for 50-Hz motor
 VI0 = V/sqrt(3) ;
 X10 = 0.30;
 X20 = 0.30;
 Xm0 = 13.08;
 R1 = 0.087;
 R2 = 0.228;
```

```
%d-q parameters
 Lm = Xm0/We;
 LS = Lm + X10/We;
 LR = Lm + X20/We;
 Ra = R1;
 RaR = R2 ;
% Operating point
 for n = 1:1:120
 Wm = 2*n*pi/60;
 Wme = (poles/2)*Wm;
 Pmech = 29840;
 Tmech = Pmech/Wm;
 Te=zeros(1001,1);
 lambdaDR = (0.8 + (n-1)*0.4/40)*Lambda;
 iQ(n) = (2/3) * (2/poles) * (LR/Lm) * (Tmech/lambdaDR) ;
 iD(n) = (lambdaDR/Lm) ;
 iQm(n) = - (Lm/LR)*iQ(n);
 Ia(n) =sqrt((iD(n)^2 + iQ(n)^2)/2) ;
We(n) = Wme - (RaR/LR)*(iQ(n)/iD(n)) ;
 fe(n) = We(n)*poles/120 ;
Varms(n) = sqrt( ((Ra*iD(n)-We(n)*(LS-Lm^2/LR)*iQ(n)) ^2 +(Ra*iQ(n)+
We(n)*LS*iD(n))^2) /2) ;

end
plot (Varms,iQ)
 figure
 plot(Varms,iD)
 end
```

E

Appendix

Modeling and Simulink of SV-PWM fed PMSM

Schematic of Approaches

Table (E-1) Parameters of PMSM Model

S. No.	PMSM Parameters	Values
1.	Stator Resistance R_s	1.4 Ω
2.	d-axis Inductance (L_d)	0.0066 H
3.	q-axis Inductance (L_q)	0.0066 H
4.	Permanent Magnet Flux	0.1546
5.	Rated Speed ω_r	522 rpm
6.	No. of Poles p	6
7.	Moment of Inertia J	0.00176
8.	Damping Coefficient B	0.0003881

Block diagram of DTC:

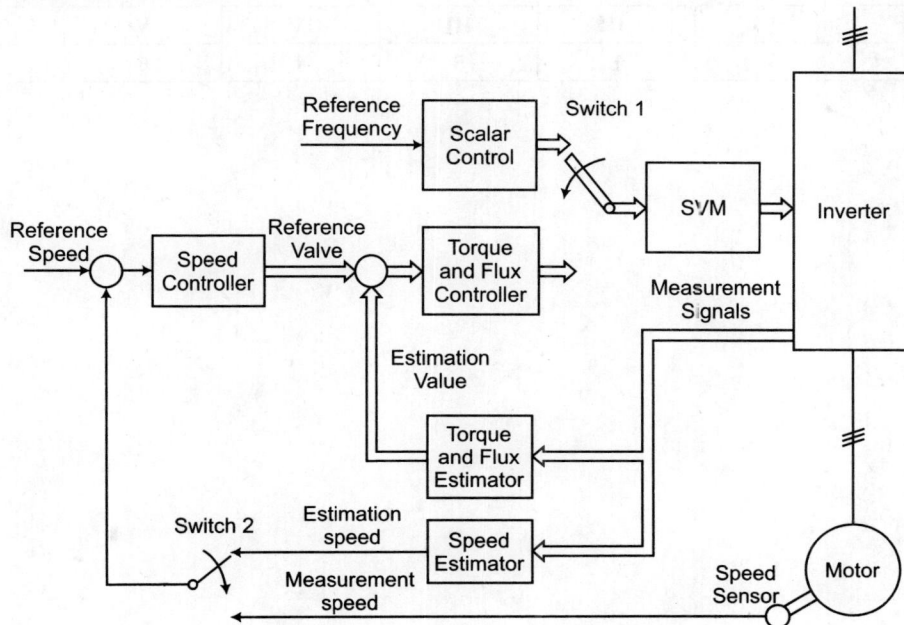

The stator voltage equation in rotor reference frame (*dq* reference frame) are given as

$$V_d = R_d I_d + \frac{d\psi_d}{dt} - \omega_r \Psi_q$$

$$V_q = R_q I_q + \frac{d\psi_q}{dt} + \omega_r$$

where R_d and R_q are the direct and quadrature axis winding resistances which are equal and be referred to as R_s in the stator resistance.

To compute the stator flux linkage in *q* and *d* axes, the current in stator and rotor are required. The permanent magnet excitation can be modeled as a constant current source, if the rotors flux along *d* axes. So the *d* axes rotor current is i_f. The *q* axes current in rotor is zero, because there is no flux along this axis in rotor, by assumption. Then the flux linkage are written

$$\Psi_q = L_q i_q$$

$$\Psi_d = L_d i_d + \Psi_f$$

Ψ_f is the flux through stator winding due to permanent magnets

$$\Psi_f = L_m i_f$$

For implementation of SVPWM first we consider the expression of vector in α-β coordinates and then following procedure is used for determining the sector. When $V_\beta > 0$, $A = 1$; when $\sqrt{3}\ V_\alpha - V_\beta > 0$, $B = 1$; when $\sqrt{3}\ V_\alpha + V_\beta < 0$, $C = 1$. Then, the sector containing the voltage vector can be decided according to $N = A + 2B + 4C$, listed in Table & Figure shows the corresponding model.

Table of Sector containing the volt. Vector versus N

Sector	I	II	III	IV	V	VI
N	3	1	5	4	6	2

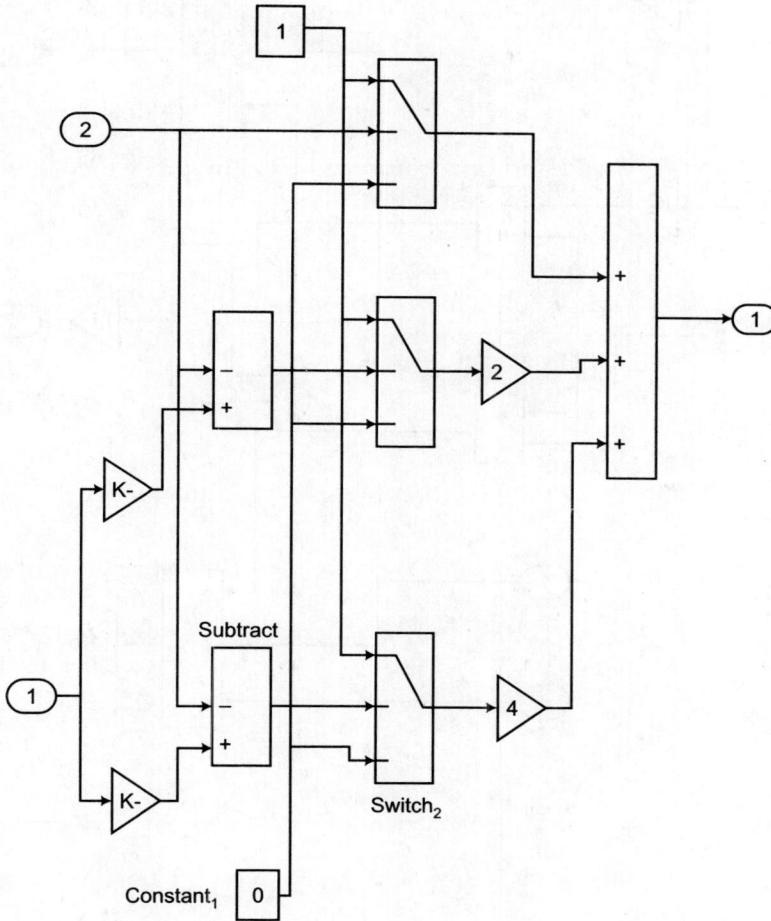

where $Z = \dfrac{T\,(-3V_\alpha + V_\beta)}{2V_{DC}}$, $Y = \dfrac{T\,(3V_\alpha + V_\beta)}{2V_{DC}}$, $X = 2T\left[\dfrac{(V_\beta)}{2V_{DC}}\right]$. The sum of T_1 and T_m must be

smaller than or equal to T (PWM period). The over saturation state must be judged: if

$T_1 + T_m > T$, take, $T_m = T_l\left[\dfrac{T}{(T_l + T_m)}\right] T_l = T_m\left[\dfrac{T}{(T_l + T_m)}\right]$.

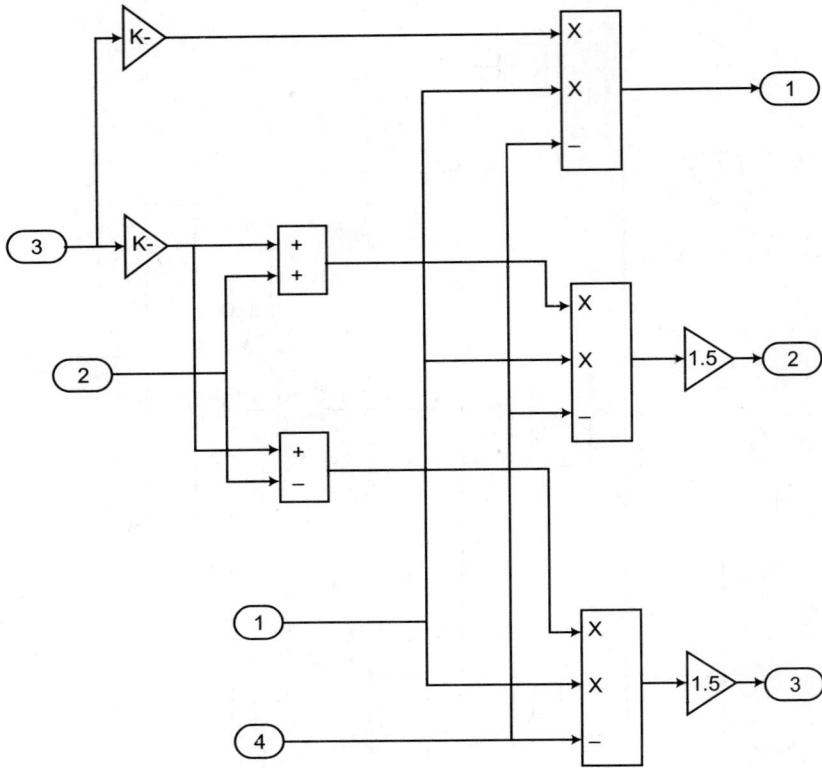

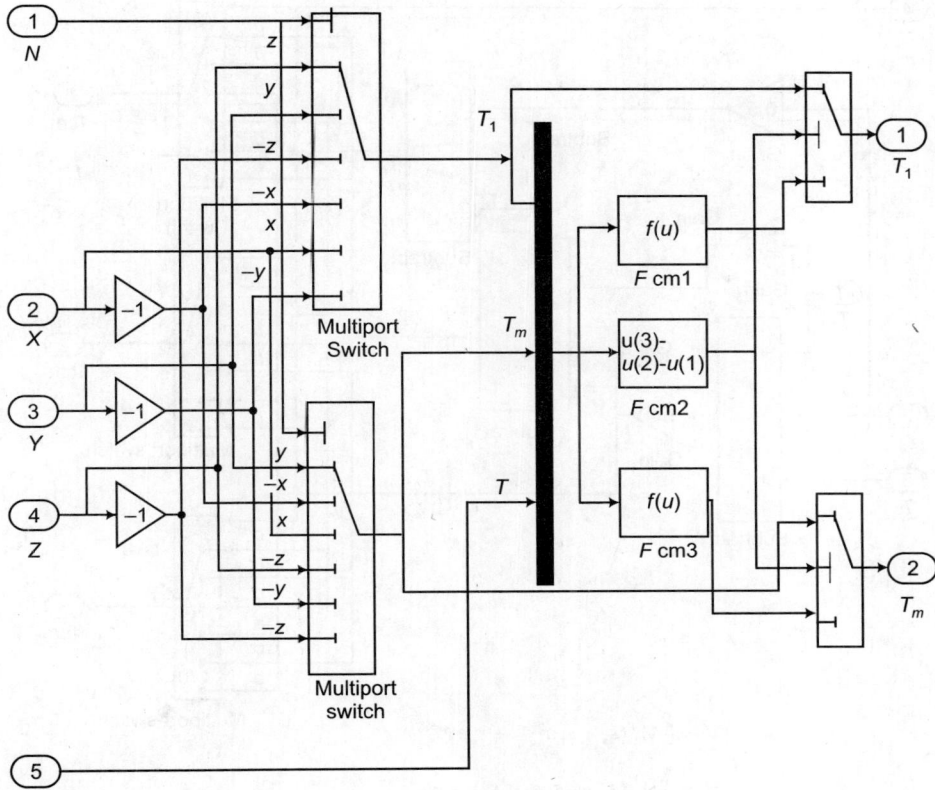

N	1	2	3	4	5	6
T_1	Z	Y	-Z	-X	X	-Y
T_m	Y	-X	X	Z	-Y	-Z

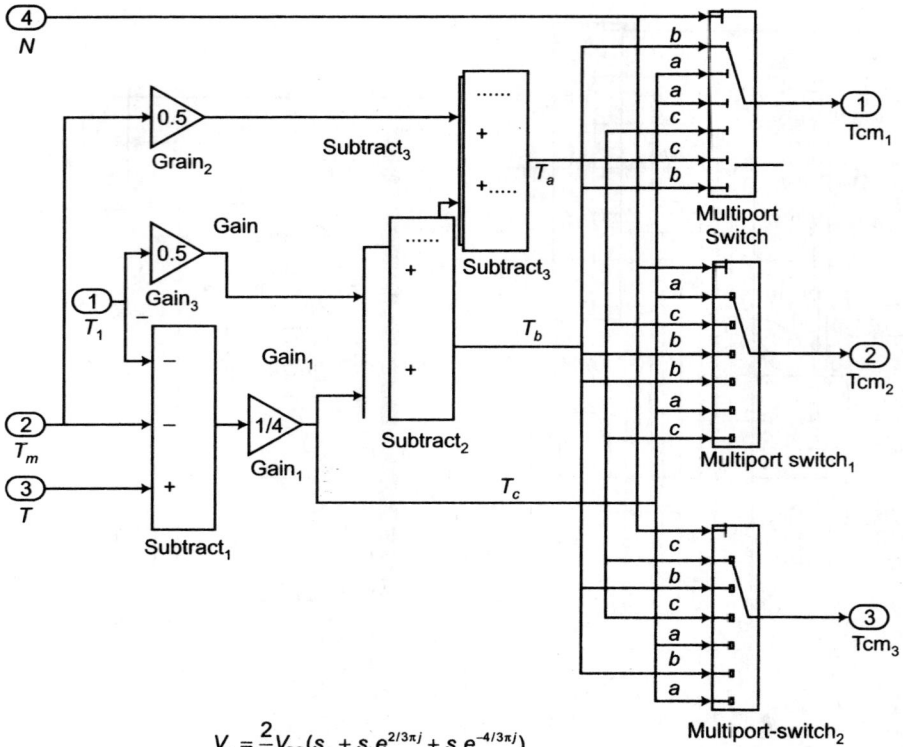

$$V_s = \frac{2}{3} V_{DC}(s_a + s_b e^{2/3\pi j} + s_c e^{-4/3\pi j})$$

$$V_s = \frac{2}{3} V_{DC}(s_a + as_b + a^2 s_c), a = e^{j2/3\pi}$$

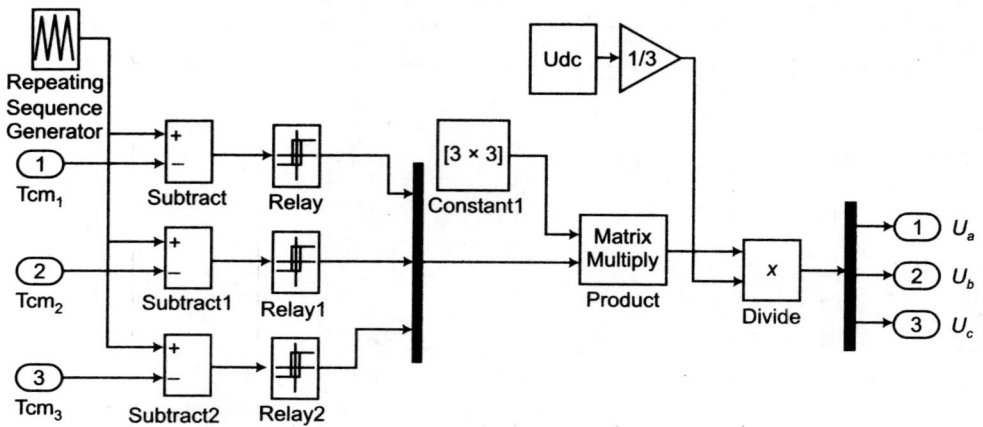

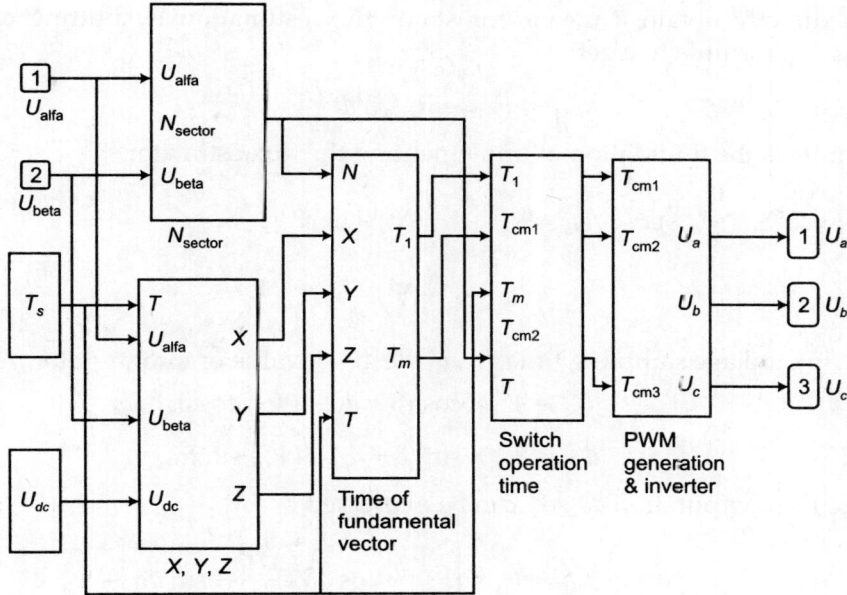

$$\begin{bmatrix} f_d \\ f_q \end{bmatrix} = \begin{bmatrix} 2/3 & -1/3 & -1/3 \\ 0 & -1/3 & 1/3 \end{bmatrix} \begin{bmatrix} f_a \\ f_b \\ f_c \end{bmatrix}$$

$$\Psi_d = \int (V_d - R_S I_S)\, dt$$

$$\Psi_q = \int (V_q - R_S I_q)\, dt$$

The flux linkage phasor is given by

$$\Psi_S = \sqrt{\Psi_{d^2} + \Psi_{q^2}}$$

$$\delta = \tan^{-1} \frac{\Psi_d}{\Psi_q}$$

The electromagnetic torque can be estimated with

$$T_e = \frac{3}{2}\, (\Psi_d I_q - \Psi_q I_d)$$

The current $i_{\alpha\beta}$ is continues the signal to the flux estimator. In this block also enters the VSI voltage vector transformed to the $\alpha\beta$-stationary reference frame. The voltage $V_{\alpha\beta}$ is calculated as in The dq-volatge equation's with zero components left is

$$V_{dq} = R_s I_{dq} + \omega_r \begin{bmatrix} -\lambda_q \\ \lambda_d \end{bmatrix} + \frac{d\lambda_{dq}}{dt}$$

One can directly obtain a means for stator flux estimation by putting $\omega_r = 0$ and rearranging this formula, we get

$$\lambda_{\alpha\beta} = \int (V_{\alpha\beta} - R_s I_{\alpha\beta})\, dt$$

This formula is the foundation of implementing the flux estimator.

$$U_d = R_S i_d + \frac{d}{dt}\Psi_d - \omega_r \Psi_q$$

$$U_q = R_S i_q + \frac{d}{dt}\Psi_q - \omega_r \Psi_d$$

where dt is flux linkage sampling time Ψ_{sr} is the given value of torque stator flux linkage

$$d_{\Psi d} = \Psi_{sr} \times \cos(\delta + d_\delta) - (\Psi_{sr})\cos\delta$$

$$d_{\Psi q} = \Psi_{sr} \times \sin(\delta + d_\delta) - (\Psi_{sr})\sin\delta$$

To reduce the computation $d_{\Psi d}, d_{\Psi q}$ can be expressed

$$d_{\Psi d} = \Psi_{sr} \times \left[\frac{\Psi_d}{(\Psi_{sr})}\cos d\delta - \frac{\Psi_q}{(\Psi_{sr})}\sin d\delta \right] - \Psi_d$$

$$d_{\Psi q} = \Psi_{sr} \times \left[\frac{\Psi_q}{(\Psi_{sr})}\cos d\delta - \frac{\Psi_d}{(\Psi_{sr})}\sin d\delta \right] - \Psi_q$$

The stator voltage component U_α, U_β is expressed in terms of U_d, U_q as

$$\begin{bmatrix} U_\alpha \\ U_\beta \end{bmatrix} = \begin{bmatrix} \cos\theta & -\sin\theta \\ \sin\theta & \cos\theta \end{bmatrix} \begin{bmatrix} U_d \\ U_q \end{bmatrix}$$

After $dq\theta/\alpha\beta$ transformation, the α-axis, β-axis voltage component of stator voltage (U_α, U_β) will be a input of SVPWM inverter to generate three-phase sinusoidal voltages fed to PMSM drive.

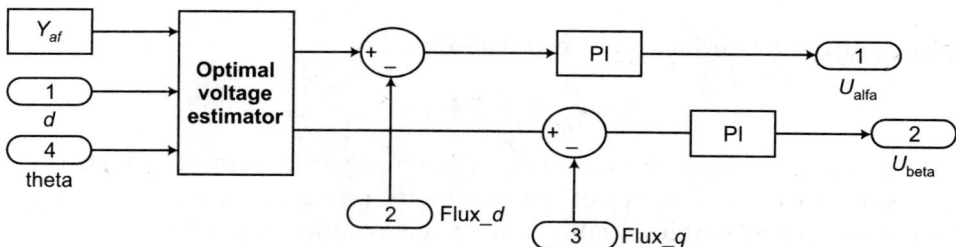

Fig. E-1 Optimal voltage estimator.

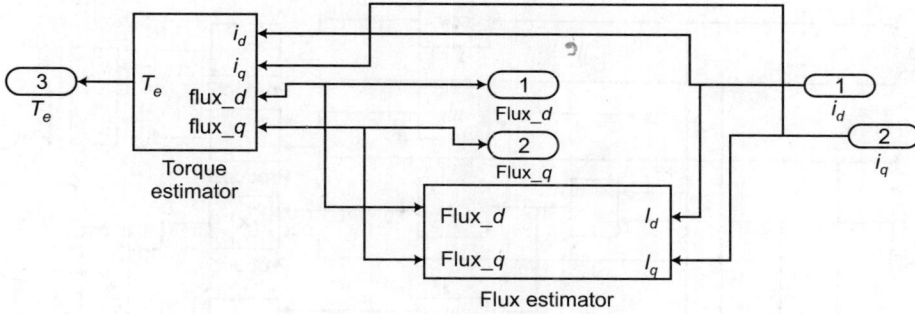

Fig. E-2 Torque and flux estimator.

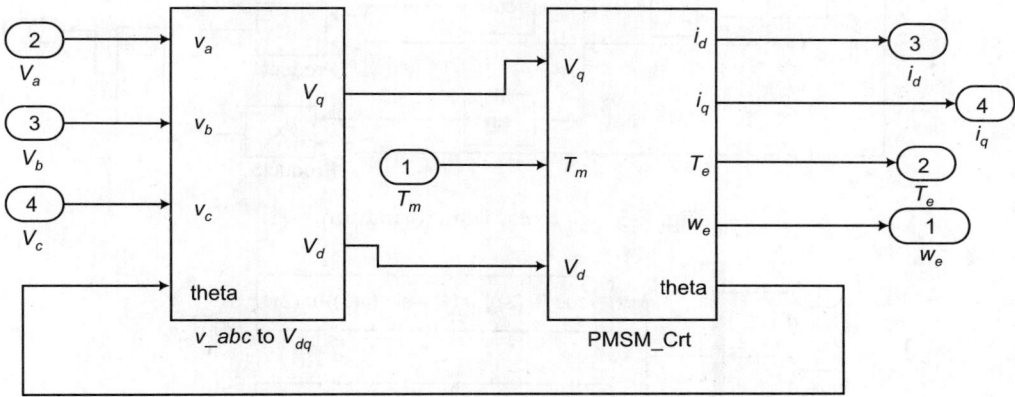

Fig. E-3 PMSM motor model.

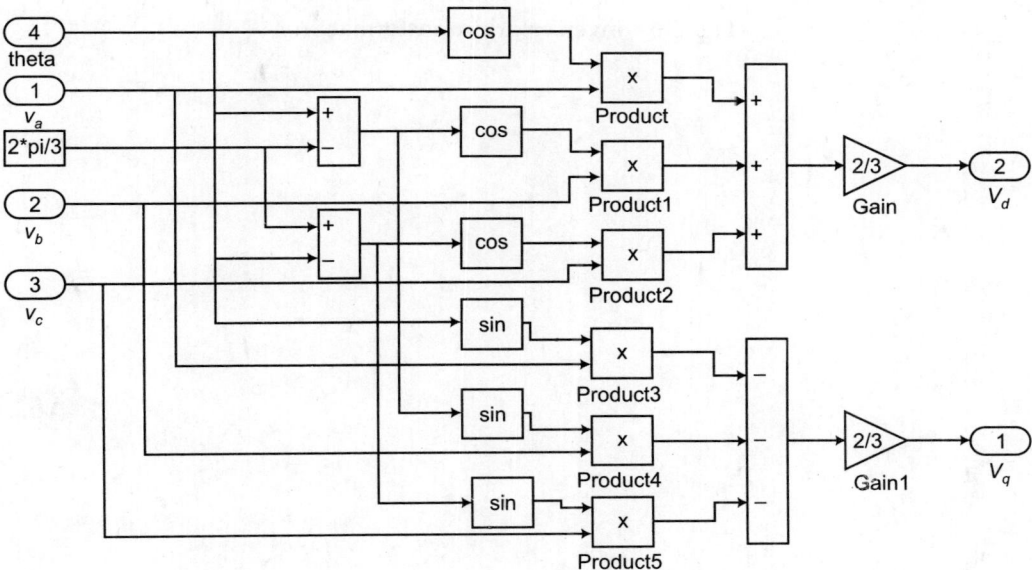

Fig. E-4 *abc* to *dq* transformation.

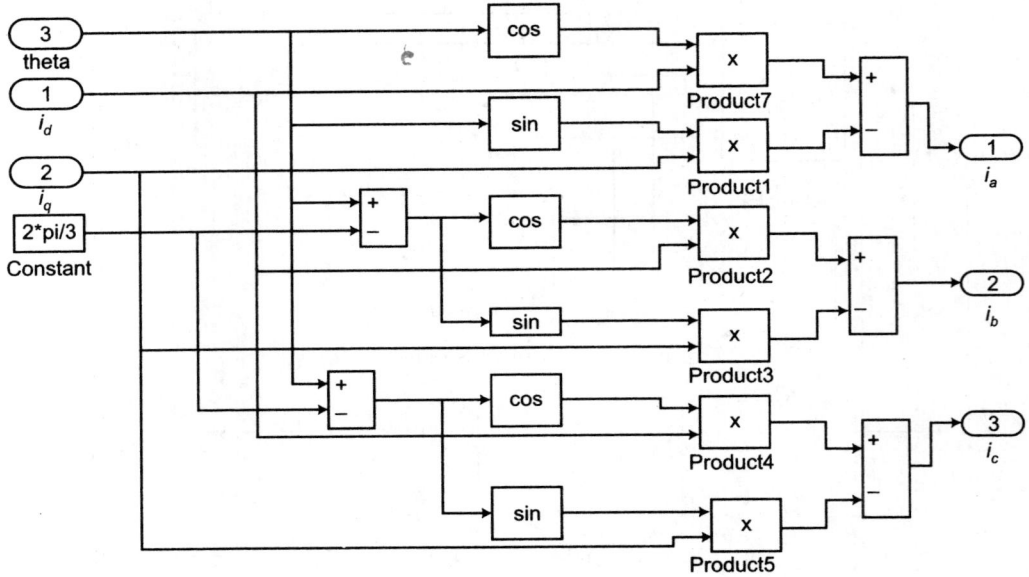

Fig. E-5   *dq* to *abc* transformation.

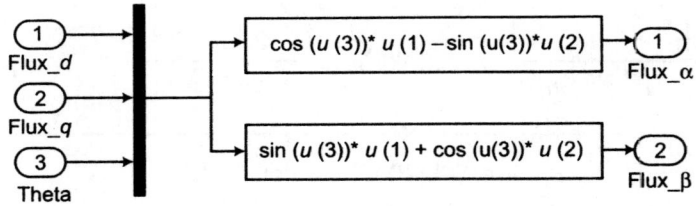

Fig. E-6   Inverse parks transformation.

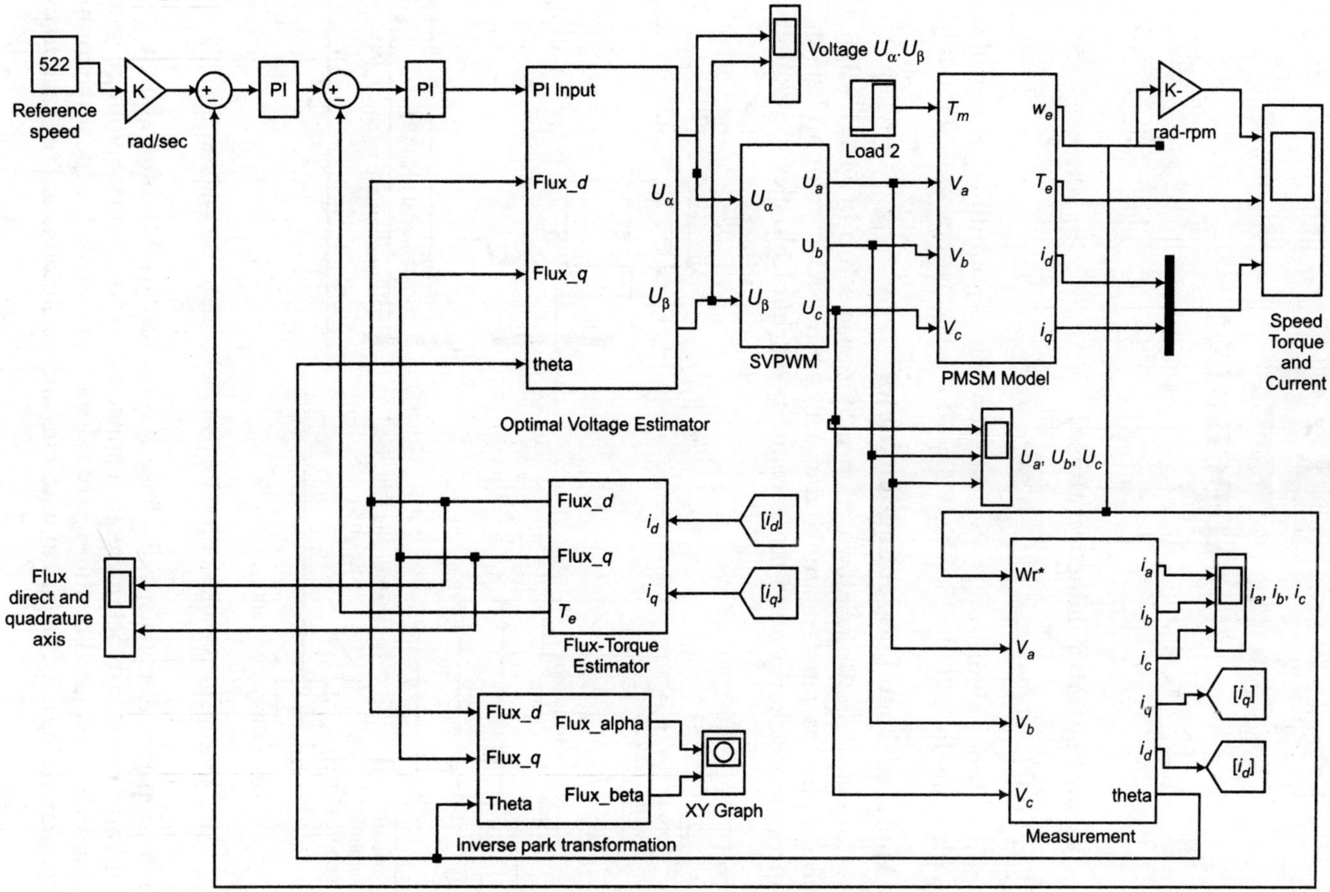

Fig. E-7 Simulink of direct torque control of PMSM.

F

Appendix

Simulink of Vector Control of Induction Motor

INTRODUCTION

To verify the developed algorithm, it was tested on Matlab/Simulink, powerful simulation software with many inbuilt blocks which is used in making a proposed model.

Simulink Model of Open Loop Induction Motor

A model of a open loop 3-phase induction motor was setup in MATLAB/SIMULINK in which controlling parameters are fixed or set by a user. The rotor and stator currents, speed, electromagnetic torque and the Torque-Speed characteristics were observed.

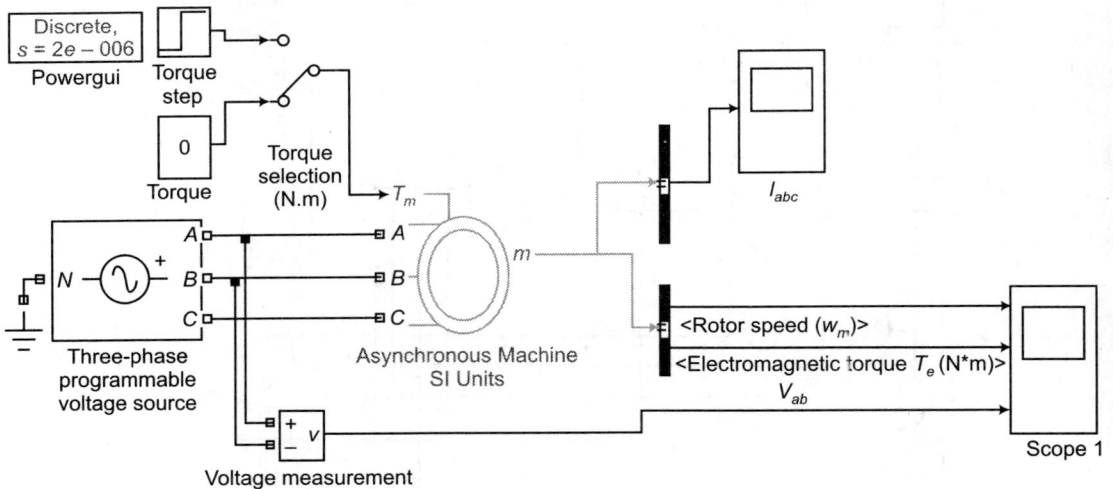

Fig. F-1   Simulink open loop model of induction motor.

The open loop model consist a 3-phase Asynchronous Machine shown in Fig. F-1 which is already existing block, directly connected to the 3-phase voltage source. The mode of operation is dictated by the sign of the mechanical torque (positive for motoring, negative for generating). All electrical variables and parameters are referred to the stator.

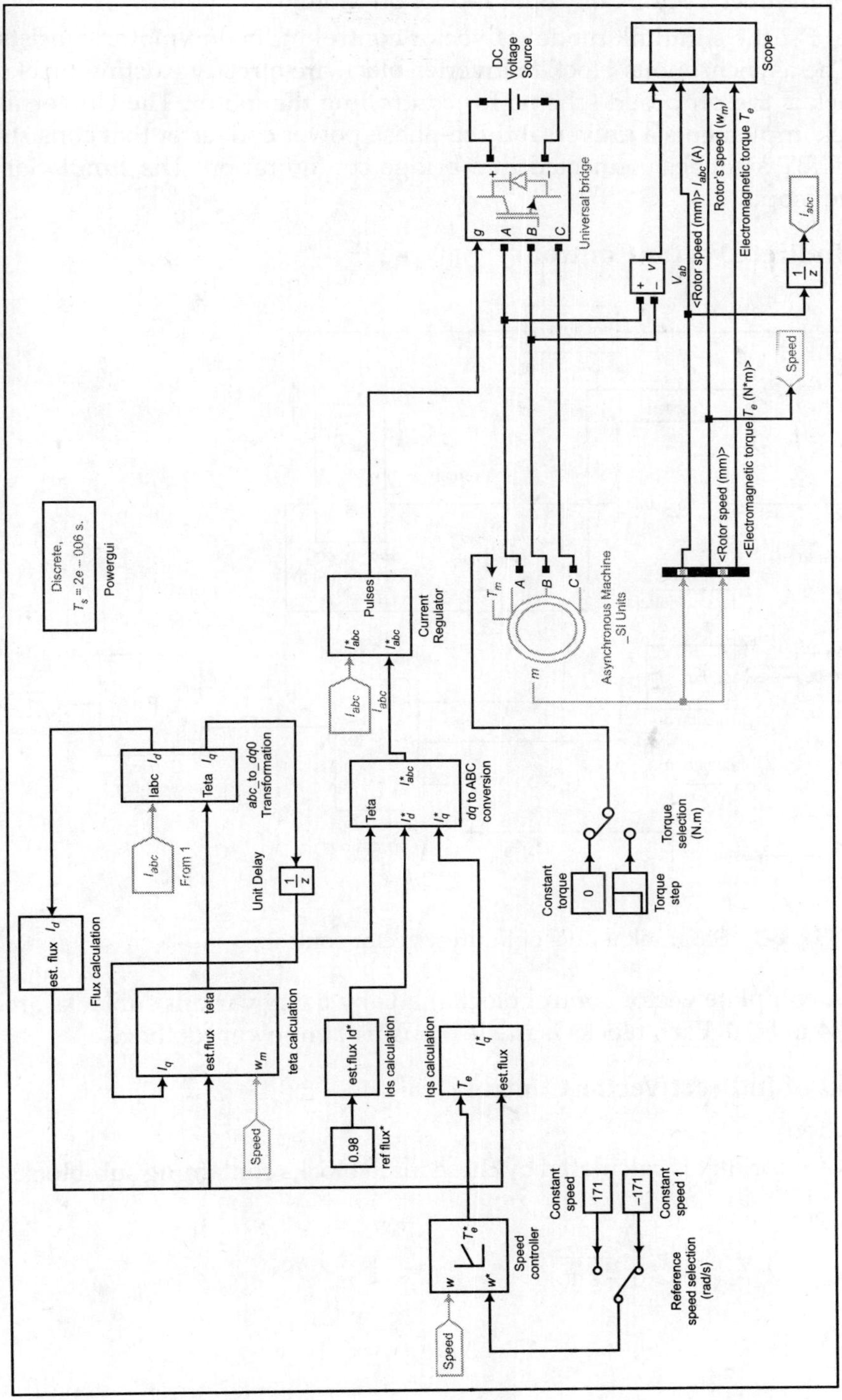

Fig. F-2 Simulation model of indirect vector of control induction motor.

Simulink Model of Indirect Vector Control of Induction Motor

As detailed in Fig. F-2 the simulink model of vector control induction motor consists three main blocks. The asynchronous block & inverter block are already existing blocks & vector control block is the proposed scheme for controlling the motor. The Universal Bridge/inverter block implements a universal three-phase power converter that consists of up to six power IGBT switches connected in a bridge configuration. The simulation results display by the scope.

Simulink Block of Indirect Vector Control

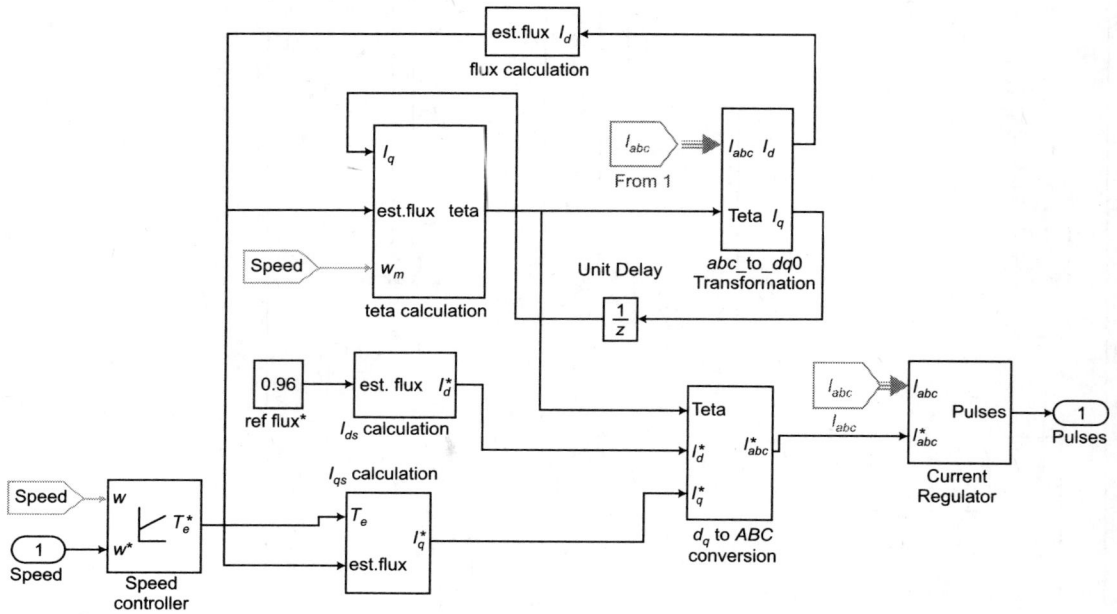

Fig. F-3   Simulink model of indirect vector control.

The Fig. F-3 shows, complete vector control block made by using various sublocks are represented in Fig. F-4 to F-10. Each blocks is made by using simpower toolbox.

Simulink Sub-Blocks of Indirect Vector Control

ψ_{est} (est. flux) Calculation

The estimated value of rotor flux is calculated by these main blocks containing sub-blocks using equation.

$$\psi_{est.flux} = \frac{L_m I_{ds}}{1 + s.\tau_r}$$

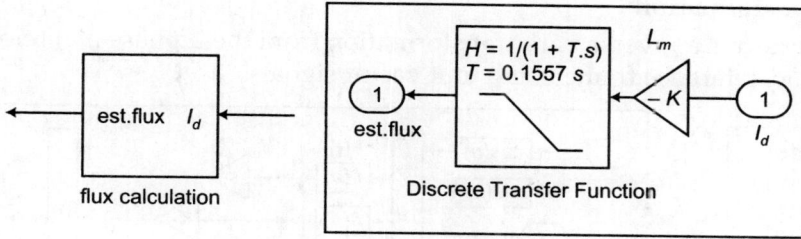

Fig. F-4 ψ_{est} (est.flux) calculation.

I_{ds} Calculation

The I_{ds}^* is calculated using reference value of rotor flux. Equation $I_{ds}^* = \dfrac{|\psi_r^*|}{L_m}$

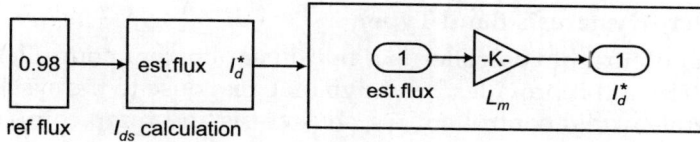

Fig. F-5 I_{ds} calculation.

I_{qs} Calculation

The I_{qs}^* is calculated using equation $I_{qs}^* = \dfrac{2}{3}\dfrac{2}{p}\dfrac{L_r}{L_m}\dfrac{T_e}{\psi_{est,flux}}$

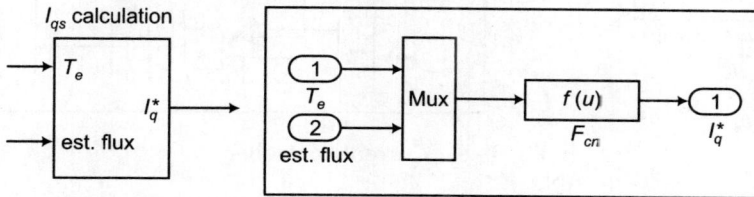

Fig. F-6 I_{ds} calculation.

Abc to dq Transformation

These blocks perform a park transformation from the 3-phase (*abc*) reference frame to 2-phase (*dq*) reference frame using unit vector signals

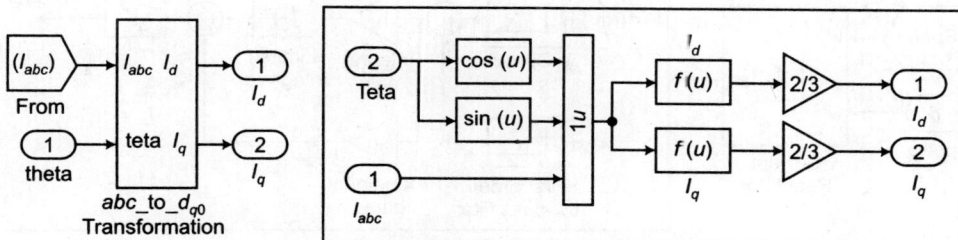

Fig. F-7 Abc to d_{q0} transformation.

dq to *abc* Transformation

This block performs a inverse park transformation from the 2-phase (*dq*) reference frame to 3-phase (*abc*) reference frame using unit vector signals.

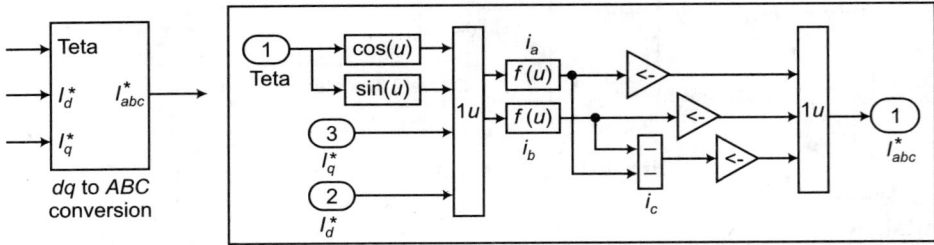

Fig. F-8   *dq* to *abc* transformation.

Current Controller (Hysteresis Band Type)

The hysteresis band current controller is a non-linear current controller does not need the load information and it provides good dynamic response to the system had given an overview of several current controllers, e.g., hysteresis and ramp comparison controller in greater depth.

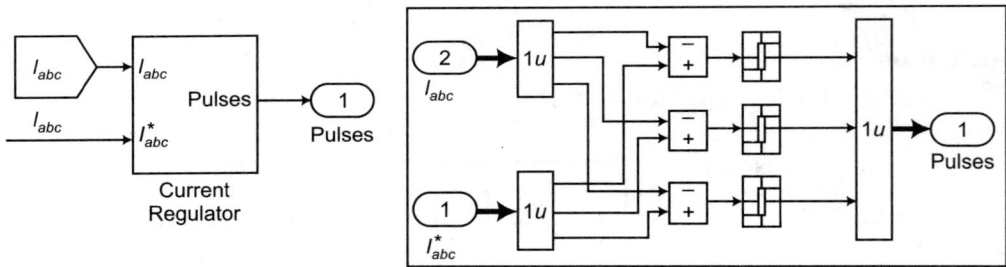

Fig. F-.9   Current controller.

Speed Controller (PI Controller)

The proportional plus integral speed controller is used to process the speed error between the reference speed & filtered speed feedback signal and is implemented using Simulink blocks.

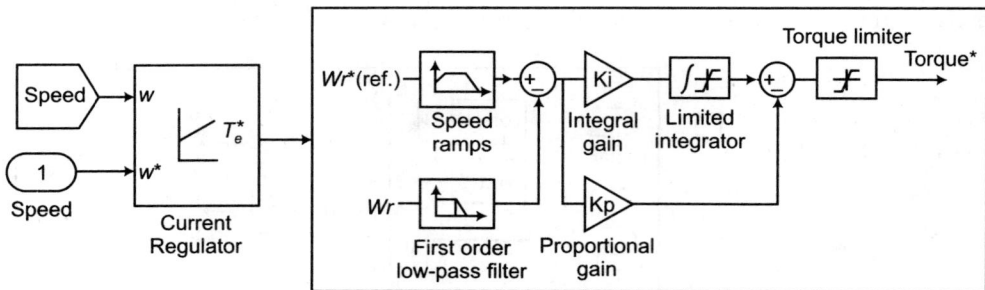

Fig. F-10   PI Speed controller.

Important Questions

1. Describe the working principle and operation of three-phase semi-controlled converter supplied to separately excited DC motor drive. Draw the O/P voltage and current waveforms. Obtain the expression of angular velocity in terms of torque with torque-speed characteristics.

2. Discuss the advantages and disadvantages of converter control of separately excited DC motor.

3. Draw the power circuit of single-phase full-wave semi-converter fed DC motor and draw relevant waveforms for O/P voltage, load current, supply current and gate signals.

4. Explain single-phase fully controlled converter giving the mathematical expression of average O/P voltage and relevant waveforms.

5. Draw and explain the speed torque characteristics of a separately exited DC shunt motor giving its area of application.

6. Write a short note on thyristor controlled DC drives.

7. Explain the concepts of constant torque control and constant power control.

8. Write a short note on single-phase fully controlled converter.

9. Describe multi-quadrant operations of DC motor drive by dual converters. Also suggest the closed loop scheme with block diagram.

10. What do you mean by "TRC" and "CLC" in chopper operation? Describe type-E configuration in all respects to control of DC motor drive for four-quadrant drive operations.

11. Write short notes:

    (1) Chopper fed DC motor for two-quadrant operation.

    (2) Regenerative braking for DC series and three-phase induction motor.

12. Draw the power circuit of two-quadrant chopper fed DC motor drive and draws the relevant waveforms considering continuous current operation.

13. Explain the operation of a closed loop speed control of DC motor drive with inner current control loop.

14. Write a short note on four-quadrant control of DC motor by dual converters.

15. State and explain the important features of various braking methods of DC motor.

16. Explain the different modes of operation of a multi-quadrant chopper controlled DC drive.

17. Make a list of advantages offered by DC chopper drives over line commutated converter controlled DC drives.

18. Explain how the speeds control of a DC motor is achieved illustrating the triggering circuit of the thyristor.

19. Discuss in delta counter current and dynamic braking operations of DC shunt motors.

20. What are the advantages of electric braking over mechanical braking of DC motor? Explain with proper circuit diagram speed torque characteristic of a DC motor under dynamic braking, for the following types:

    (i) Separately excited DC motor.

    (ii) Series motor.

21. With a neat diagram, explain the operation of a DC drive in all four-quadrants when fed by a single-phase dual converter with necessary waveform and characteristics.

22. Write a short note on closed loop speed control operation of DC drives.

23. A 240 V, 950 rpm., 14 Amp. Separately excited DC motor has armature circuit resistance and inductance of 2 ohm and 150 mH respectively. It is fed from a single-phase half-controlled rectifier with an AC source voltage of 250 V, 50 Hz. Calculate:

    (i) Motor torque for firing angle $\alpha = 60°$ and speed = 600 rpm.

    (ii) Motor speed for $\alpha = 60°$ and $T = 20$ Nm.

24. A 220 V, 1500 rpm, 50 A separately excited DC motor with armature resistance of 0.5 ohm is fed from a three-phase fully controlled converter. The input to the converter is three-phase, 230 V, 50 Hz AC supply. Determine the firing angle, when the motor is running 1200 rpm and rated torque.

25. A 200 V, 875 rpm, 150 A separately exited DC motor has an armature resistance of 0.06 ohm. It is fed from a single-phase fully controlled rectifier with an AC source = voltage of 220 V, 50 Hz. Assuming continuous conduction, calculate:

    (i) Firing angle for rated motor torque and 750 rpm.

    (ii) Motor speed for $\alpha = 60°$ and rated torque.

26. A 400 V, 750 rpm, 70 A DC shunt motor has an armature resistance of 0.3 ohm when running under rated conditions, The motor is to braked by plugging with armature current limited to 90 A. what external resistance should be connected in series with the armature ? Calculate the initial braking torque and its value when the speed has fallen to 300 rpm.

27. A single-phase, half controlled converter is fed from 120 V. rpm, 60 Hz supply and provides a variable DC voltage at the terminals of a DC motor. The thyristor is triggered continuously by a DC signal. The resistance of armature circuit is 10 ohm and because of fixed motor excitation and high inertia, the motor speed is considered constant so that the back emf is 60 V. Find the average value of the armature current neglecting armature inductance.

28. A 230 V, 960 rpm and 150 A separately excited DC motor has an armature resistance of 0.025 ohm. The motor is fed from a chopper which provides both motoring and braking operations. The source has a voltage of 230 V.

     Assume continuous conduction.

29. A 2 kW, 230V, 10 A, 1500 rpm separately excited DC drive has an armature resistance and inductance of 2 ohms and 32 H respectively. A chopper controls the motor with a frequency of 500 Hz and input voltage 230 volts. The motor is driving a load whose torque is proportional to speed. At duty ratio of 0.9, the motor runs at 1260 rpm. What will be the value of duty ratio at 800 rpm?

30. A 230 V, 960 rpm and 200 A. separately excited DC motor has an armature resistance of 0.02 ohm. The motor is fed from a chopper which provides both motoring and braking operation. The source has a voltage of 230 volt assuming continuous conduction.

     (i) Calculate duty ratio of chopper for motoring operation at rated torque and 350 rpm.

     (ii) Calculate duty ratio of chopper for braking operation at rated torque and 350 rpm.

31. Describe the PWM control of a three-phase induction motor in detail. Draw its waveform, speed torque characteristics and obtain the expression of maximum torque. Also suggest the closed loop scheme with block diagram.

32. Write short notes on variable frequency control of induction motor by VSI and CSI schemes.

33. Variable frequency control of induction motor is more efficient then stator voltage control. Why?

34. Discuss the V/f scheme of speed control of induction motor fed from a voltage source inverter for above and below base speed control. Also draw the speed torque characteristics.

35. Write short notes on PWM controlled induction motor drive.

36. Give reasons for the following:

     (i) Stator voltage control is suitable for speed control of induction motor in fan and pump drives.

     (ii) Stator voltage control is an inefficient method of speed control.

37. Explain variable voltage variable frequency control of an induction motor giving its speed torque characteristics under different modes of operation.

38. Write short notes on VSI and CSI fed induction motor drives?

39. Starting from fundamentals prove that torque developed by the induction motor is proportional to square of the supply voltage.

40. Describe static rotor resistance control of three-phase induction motor from rotor side with closed loop scheme and draw the block diagram.

41. Describe slip power recovery static Scherbius drive with complete circuit diagram. Discuss their relative advantages and disadvantages over other methods. Also give field of application of the above method.

42. Write a short note on static Kramer drive.
43. State and explain with block diagram the slip power recovery scheme (static Scherbius control) for speed control of a three-phase induction motor.
44. Draw the block diagram for stator voltage control of a three-phase induction motor drive with close loop scheme and also draw the speed torque characteristics for the same.
45. Write a short note on static Kramer drive.
46. Explain the operation of static Kramer drive with relevant block diagram and mathematical expression.
47. What are the advantages of static rotor resistance control over conventional method of rotor resistance control?
48. Write a short note on slip power recovery scheme.
49. A Y- connected SCIM has the following rating and parameters: 400 V, 50 Hz, 4-pole, 1370 rpm, $R_s = 2$ ohms, $R'_r = 3$ ohms, $X_s = X'_r = 3.5$ ohms motor is controlled by a VSI at constant V/f ratio. Inverter allows frequency variation from 10 to 50 Hz.
    (i) Obtain a pole between the breakdown torque and frequency.
    (ii) Calculate starting torque and current of this drive as a ratio of their values when motor is started at rated voltage and frequency.
50. A 440 V, 50 Hz, 960 rpm, 6-pole, Y-connected, three-phase slip ring induction motor has the following parameter referred to the stator.
    $R_s = 0.1$ ohm, $R'_r = 0.08$ ohm, $X_s = 0.3$ ohm, $X'_r = 0.4$ ohm.
    The stator to rotor turns ratio is 2.
    Motor is controlled by static-Scherbius drive. Drive is deigned for a speed range of 25% below the synchronous speed. Maximum value of firing angle is 165°. Calculate:
    (i) Transformer turn ratio
    (ii) Torque for a speed of 780 rpm and $\alpha = 150°$
    (iii) Firing angle for half the rated motor torque and speed of 800 rpm.
    DC link inductor has a resistance of 0.02 ohm.
51. An inverter supplies a 6-pole, three-phase cage induction motor rated at 415 V, 50 Hz. Determine the approximate voltage required of the inverter for motor speed 600/800/1500/1800 rpm.
52. A 440 V, 50 Hz, 6-pole, star connected wound rotor induction motor has the following parameters referred to stator:
    $R_s = 0.5$ ohm, $R'_r = 0.4$ ohm
    $X_s = X_r = 1.2$ ohm, $X_m = 50$ ohm
    Stator to rotor turn ratio is 3.5. Motor speed is controlled by static rotor resistance control. External resistance is chosen such that the breakdown torque is produced at standstill for a duty ratio of zero. Calculate the value of external resistance.

53. A 2.8 kW, 400 V, 50 Hz, 4-pole, 1370 rpm, delta connected squirrel cage induction motor has the following parameters referred to the stator:

    R_s = 2 ohm, R_s' = 5 ohm, X_S = X_r' = 5 ohm, X_m = 80 ohm.

    Motor speed is controlled by stator voltage control. When driving a fan load it runs at rated speed at rated voltage. Calculate:

    (i) Motor terminal voltage, current

    (ii) Torque at 1200 rpm.

54. A three-phase, 400 V, 15 kW, 1440 rpm, 50 Hz, star connected induction motor has rotor leakage impedance of 0.4 + j 1.6 ohm. Stator leakage impedance and rotational losses are assumed negligible. If this motor is energized from 120 Hz, 400 V, three-phase source, then calculate:

    (i) Motor speed at rated load

    (ii) Slip at maximum torque

    (iii) Maximum torque

55. Explain the operations of self-control synchronous motor by cycloconverter with circuit diagram.

56. Discuss the load commutated CSI fed synchronous motor operation with waveforms speed torque characteristics. Also give the field of applications.

57. Write short notes on separate and self-control of synchronous motor.

58. Describe self-control of synchronous motor drive and enlist its advantages.

59. Discuss the load commuted CSI fed synchronous motor operation with relevant waveforms and speed torque characteristics.

60. Discuss the load commutated CSI fed synchronous motor drive with block diagram and relevant waveforms?

61. Write a short note on closed loop operation of synchronous motor drive.

62. Differentiate true synchronous mode and self-control mode for variable frequency control of synchronous motor drive?

63. With the help of a block diagram explain closed loop speed control scheme of synchronous motor drive.

64. Write a short note on applications and advantages of synchronous motor drive.

65. Describe the open loop and closed loop methods of speed control of a synchronous motor using VSI.

References

1. Dubey, G.K., *Fundamentals of Electrical Drives*. New Delhi, Narosa Publishing House, 2009.
2. Bimbhra, P.S., *Power Electronics*. New Delhi, Khanna Publishers, 2006.
3. Bose, B.K., Power electronics and motor drives recent technology advances, *Proceedings of the IEEE International Symposium on Industrial Electronics,* IEEE, 2002.
4. Rashid, M.H., *Power Electronics,* Prentice-Hall of India, New Delhi, 1993.
5. Matlab and Simulink Version 2009a, the Mathsworks Inc, USA.
6. Zuo Z. Liu, Fang L. Luo, and Muhammad H. Rasid, "High Performance nonlinear MIMO field weakening controller of a Separately Excited DC Motor" *Electric Power Systems Research,*Vol. 55, Issue 3, Sept. 2000, pp. 157-164.
7. Simulink, Model-based and system-based design, Using Simulink, MathWorks Inc., Natick, MA, 2000.
8. Sim Power Systems for use with Simulink, user's guide, MathWorks Inc., Natick, MA, 2002.
9. M.D. Singh, K.B. Khanchandani, *Handbook of Power Electronics,* second edition, Tata Mc-Graw-Hill.
10. M. Gautier, P.M. Robet and C. Bergmann," Modeling and Simulation of DC-Motor Chopper Drive for Robots".
11. Franc Mihalic and Dejan Kos "Randomized PWM for conductive EMI reduction in DC-DC choppers,"*HAIT Journal of Science and Engineering B*, Vol. 2, Issues 5-6, pp.
12. Gopal, M., *Control Systems, Principles and Design*. New Delhi, Tata McGraw-Hill Publishing Company Ltd, 2008.
13. Mohan, Ned, Electrical Drives–An Integrated Approach. Minneapolis, MNPERE, 2003.
14. Moleykutty George, Speed Control of Separately Excited DC Motor, *American Journal of Applied Sciences* 5 (3): 227-233, 2008 ISSN 1546-9239 © 2008 Science Publications.
15. R. Krishnan, *Handbook of 'Electric Motor Drives'* pp. (133-134) Pearson Education.
16. Wester, G.W. & Middle Brook, R.D. "Low- frequency characterization of switched DC-DC converters", IEEE Transactions on Aerospace and Electronic Systems, Vol. AES- 9, No. 3, pp. 376-385, 1973.
17. Rakesh Singh Lodhi, Analysis & Design of PID Controller for Chopper-fed Separately Excited DC Motor, National Conference of ECOMM-11.

18. Bose B.K., Power electronics and motor drives recent technology advances, Proceedings of the IEEE International Symposium on Industrial Electronics, IEEE, 2002, pp. 22-25.

19. Brijesh Singh, Intelligent PI Controller for Speed Control of D.C. Motor, *International Journal of Electronic Engineering Research* ISSN 0975 – 6450 Vol. 2 No. 1 (2010) pp. 87–100 © Research India Publications.

20. S.N. Sivanandam and S.N. Deepa, *Handbook of 'Control Systems Engineering using MATLAB'* pp. (91-92) Vikash Publishing House Pvt Ltd.

21. Rakesh Singh Lodhi and H.K. Verma "Analysis & Design of PID Controller for Chopper-fed Separately Excited DC Motor" NCRTEE, 2011.

22. Brijesh Singh, Intelligent PI Controller for Speed Control of D.C. Motor, *International Journal of Electronic Engineering Research* ISSN 0975 – 6450 Vol. 2 No. 1 (2010) pp. 87–100 © Research India Publications.

23. S.J. Chapman, *Electric Machinery Fundamentals*, 3rd Edition WCB/McGraw-Hill, New York 1995.

24. Mohan, Undeland, Robbins, "Power Electronic", second edition, Wiley, 1989.

25. N.P Quang and J.A. Dittrich, "*Vector Control of 3-phase AC Machines*, Springer, 2008.

26. Bruno Francois and Philippe Degobert, "*Vector Control of Induction Machines*" Springer February 9, 2012.

27. Ali S. Ba-thuya, Rahul Khopkar, Kexin Wei Hamid and A. Toliyat, "Single-phase induction motor drives — A Literature survey", Electric Machines & Power Electronics Laboratory Texas.

28. R. Krishnan, "*Electric Motor Drives-Modelling, Analysis & Control*", 2001 Prentice Hall.

29. F. Blaschke, "The principle of field orientation as applied to new transvector close loop system for rotating field machine", 1972.

30. P. Vas, "*The Control of A.C Machine*" Oxford Univ. 1990.

31. Krause P.C, "*Analysis of Electric Machinery*" McGraw-Hill, New York, 1986.

32. B.K.Bose, "*Modern Power Electronics and AC Drives*", Pearson Education, 4th Edition, 2004.

33. Aung Zaw Latt and Ni Niwin,"Variable speed drives of single phase induction motor use in frequency control methods", International Conference on Education Technology & Computer, 2009.

34. S. Jeevananthan, "Matlab/Simulink — A Tool for Power Electronic Circuits", Pondicherry Engineering College, Pondicherry.

35. Krause P.C and C.S. Thomas "Simulation of symmetrical induction machinery", IEEE Trans. on Power Apparatus & Systems, Vol. 84, No. 11, 1965, pp. 1038- 1053.

36. M.P. Kazmierkowaski and L. Malesani, "PWM current control techniques of voltage source converters-A Survey", IEEE. Trans. on Industrial Electronics, Oct. 1998. Vol. 45.

37. D.M. Brod and W. Novotny, "*Current control of VSI-PWM inverters*", IEEE IAS Annual Meeting, 1984.

38. Dal Y. Ohm Drivetech, *"Dynamic Model of Induction Motors for Vector Control"*, Inc., Blacksburg, Virginia.

39. V. Lakaparampil and V.T. Ranganathan, "Modelling, Simulation & Implementation of Vector Controlled Induction Motor Drive".

40. Alexandru Onea, Vasile Horga and Marcel Ratoi,"Indirect Vector Control of Induction Motor", Department of Electric Drives "Gh. Asachi", Technical University of Iaşi Bd. Mangeron. 53A, 6600, Iaşi, Romania.

41. C.M. Ong, *"Dynamic Simulation of Electric Machinery"*, Prentice Hall, New Jersey, 1998.

42. R.H Park, "Two-reaction theory of synchronous machine-generalized method of analysis Part 1", AIEE Trans, Vol. 48 , pp. 716-727 July 1929.

43. Dheeraj Joshi and Meghna Gill, "Comparison of Vector Control Techniques for Induction Motor Drives", *Indian Journal of Electrical Biomedical Eng.*, Vol. 1, 2013.

44. Bhim, B.N and B.P Singh, "Performance Analysis of a Low Cost Vector Control Induction Motor Drive", IEEE IAS Conf, pp. 789-794, 1997.

45. Mohammad Zadeh Rostami, "Analysis of indirect rotor field oriented vector control of squirrel cage induction motor drives", IEEE International Conf. 6-7 June 2012.

46. Selmon, G.R , "Modeling of induction machine for electric drives", IEEE Vol. 25, Issue 6, 1989.

47. A. Khodabakshian and K. Jamshidi, "Vector Control of Induction Motor Using UPWM Voltage Source Inverter", G. Esmeily, Isfahan Unv., Iran.

48. Luigi Malesani and Paolo Tomasin, "PWM Current Control Technique of Voltage Source Converter", Univ of Podova, Italy.

49. C.S. Sharma and Tali Nagwani, "Simulation & Analysis of PWM Inverter Fed Induction Motor Drives", Ijsetr Vol. 2, Feb 2013.

50. G. Kohlrursz, D.F. Odar and Hungavan, "Comparision of Scalar & Vector Control Strategies of Induction Motor", Journal of Industrial Chemistry Veszprem, Vol. 39, 2011

51. S. Corino and E. Romero, "How the efficiency of induction motor is measured", L.F Mantilla, IEEE.

52. Ashutosh Mishra and Prashant Choudhary, "Speed control of an induction motor by using indirect vector control method" IJETAE, Dec 2012.

53. Hasse K. "On the dynamic behavior of induction machines driven by variable frequency and voltage sources", ETZ Arch. Bd. 89, H. 4, 1968, pp. 77-81.

54. Atkinson D.J., P.P. Acarnley and J.W. Finch, "Application of estimation technique in vector controlled induction motor drives", IEEE Conference Proceedings, London, July 1990, pp. 358-363.

55. Chan C.C., Leung W. S. and C.W. Nag, "Adaptive decoupling control of induction motor drives," IEEE Transaction on Industrial Electronics, Vol. 35, No. 1, Feb. 1990, pp. 41-47.

56. I. Takahashi and T. Noguchi "A new quick response & high efficiency control strategy of induction motor", IEEE Trans. Ind, 1986.

57. Gilbert Sybille and Patrice Brunelle,"Simpower System-User Guide", Trans Energie Technology, 2001.

58. Santosh B. Kulkarni and Rajan H. Chile,"MATLAB/SIMULINK Simulation Tool for Power Systems", IJPSOE 2011.

59. Atul M. Gajare and Nitin R. Bhasme,"A Review on Speed Control Techniques of Single Phase Induction Motor", IJCTEE Vol. 2, Issue 5, October 2012.

60. Mohamayee Mohapatra and B. Chitti Babu, "Fixed and sinusoidal-band hysteresis current controller for PWM voltage source inverter with LC filter", Member IEEE, 2010.

61. Novotny, D.W. and Lipo, T.A., 1996. "Vector Control and Dynamics of AC Drives", Oxford University Press, New York.

62. Correa, M.B.R, "Field Oriented Control of a Single-Phase Induction Motor Drive", Conf. Rec. Power Electronics Specialists, PESC'98, Fukuoka, Japan, Vol. II, pp. 990-996, 1998.

63. G.J. Ritter, Budape and Kelemen, A., Imecs. M, "Vector Control of AC Drives", 1987. Texas Instrument, "Field Orientated Control of 3-Phase AC Motors".

64. S. Albert Alexander,"A Comparison of Simulation Tools for Power Electronics", ISCI 2012.

65. Abbondanti, Brennen, M.B., "Variable speed induction motor drives use electronic slip calculator based on motor voltages and currents", IEEE Trans. on Indus. Applications. IA-11 pp. 483–488, 1975.

66. K. Koga, R. Ueda and T. Sonoda,"Achievement of high performances for general-purpose inverter drive induction motor system", IEEE IAS Annual Meeting, pp. 415-425, conf. 1989.

67. A. Munoz-Garcia, T.A. Lipo and D.W. Novotny, "A new induction motor open-loop speed control capable of low frequency operation", IEEE Trans. on Industry Applications, Vol. 34, July/August, pp. 813-821.

68. T. Sukegawa, K. Kamiyama and T. Matsui, T. Okuyama, "Fully digital, vector controlled PWM-VSI fed AC drives with an inverter dead-time compensation strategy". IEEE IAS Annual Mtg., pp. 463- 469,1988.

69. D.W. Novotny and T.A. Lipo,"Vector control and dynamics of AC drives", Oxford Press, Oxford England, 1996.

70. R. de Doncker and D.W. Novotny, "The universal field oriented controller", IEEE IAS Annual Meeting, pp. 450-456 Oct. 1988.

71. X. Xu and D.W. Novotny, "Selection of the flux reference for induction machines in the field weakening region", IEEE Vol. 28, pp. 1353-1358 November/December 1992.

72. K.B. Nordin, D.W. Novotny and D.S. Zinger, "The influence of motor parameter deviations in feedforward field orientation drives systems", IEEE Trans. in Industry Applic., Vol. IA-21, No. 4, pp. 1009-1015, July/August 1985.

73. Q. Yao and D.G. Holmes, "A simple, novel method for variable-hysteresis-band current control of a three phase inverter with constant switching frequency", IEEE IAS Annual Meeting, , pp. 1122- 1129,1993.

74. A. Ansari and D.M. Deshpande, "Mathematical Model of Asynchronous Machine in MATLAB Simulink", *International Journal of Engineering Science and Technology* Vol. 2(5), pp. 1260-1267, 2010

75. R. Lee, P. Pillay and R. Harley, "D,Q reference frames for the simulation of induction motors," *FPSR Journal,* Vo1. 8. October.

76. K.H. Bayer and H. Waldmann, M. Weibelzahl, "Field-Oriented Closed-Loop Control of a Synchronous Machine with the New Transvector Control System," *Siemens* Vol. 39, pp. 220-223, 1972.

77. Kelemen A.G.J. and Ritter, M., 1987. "Indirect Vector Control of Induction Motor Drives", Budapest, Imecs.

78. Kovacs, P.K., *"Transient Phenomena in Electrical Machines"*, Elsevier Science Publihsers, Amsterdam 1984.

79. Kovacs, P.K. and Racz, L. 1959. "Transiente Vorgänge in Wechselstrommaschinen", Ungarischen Akademie der Wissenschaften, Budapest.

80. Lai, Y-S. 1999. "Modelling and Vector Control of Induction Machines — A New Unified Approach", in Conf. Rec. Power Engineering Soc. Winter Meeting, Vol. I, pp. 47-52.

81. Murata, T., Tsuchiya, T. and Takeda, I., 1990. "Vector Control for Induction Machine on the Application of Optimal Control Theory", in IEEE Trans. Ind. Electronics, Vol. 37, No. 4, pp. 282-290.

82. Popescu, M. and Navrapescu, V. 2000. "Modelling in Stationary Frame of Single and Two-phase Induction Machines Including the Effect of Iron Loss and Magnetising Flux Saturation" — in Proceedings of International Conference of Electrical Machines, ICEM 2000, 28-30 August, Espoo, Finland, Vol. I, pp. 407-411.

83. Popescu, "Analysis and Modelling of Single-phase Induction Motor with External Rotor for Domestic Applications" – in Proceedings of IEEE-IAS Annual Meeting, 8-12 October, Rome, Italy, Vol. I, pp. 463-470.

84. Slemon, G.R. "Electrical Machines for Variable-Frequency", in Proceedings of the IEEE, Vol. 82, No. 8, pp. 1123-1138, 1994.

85. R.W. Smeaton, *"Motor Application and Maintenance"* Handbook, McGraw Hill Book Co.

86. W.I. Ibrahim, M.T. Raja and Ismail, M.R. Ghazali, "Development of Variable Speed Drive for Single-phase Induction Motor Based on Frequency Control", 4th Engineering Conference (EnCon 2011), 29 Nov – 1 Dec 2011.

87. D.S. Henderson,"Variable Speed Electric Drives – Characteristics & Applications", Bulletin Adjustable Frequency Control (Inverters) Fundamentals Application Consideration, C870A.

88. A.S. Zein El-Din and A.E.El-Sabbe," A novel speed control technique for single-phase induction motor," IEEE International Conference on Power Electronics and Drive Systems, PEDS' 99, July 1999, Hong Kong.

89. Wade S., Dunnigan M.W. and Williams BW, "Modeling and simulation of induction machine vector control with rotor resistance identification"IEEE Trans Power Electronics 1997.

90. Lin, F.K. and Liaw, C.M. "Control of indirect field-oriented induaction motor drives considering the effects of dead-time and parameter variations." IEEE Trans. Indus. Electro. Vol. 40, pp. 486-495.1993.

91. M. Satheesh Kumar, P. Ramesh Babu and S. Ramprasath 'Four-quadrant comparative evaluation of classical and space vector PWM-direct torque control of VSI fed three-phase induction motor drive in MATLAB/SIMULINK environment' IEEE International Conference on Power Electronics, Drives and Energy Systems December 16-19, 2012, Bengaluru , India.

92. Salam Ibrahim Khather 'Modeling and Simulation of a PWM Rectifier Inverter Induction Motor Drive System Implementing Speed Sensor Less Direct Vector Control' 9th International Multi-Conference and System, Signal and Device 2012.

93. V. Siva Krishna Rao and K. Vijaya Bhaskar Reddy, 'Modeling and Simulation of Modified Sine PWM VSI Fed Induction Motor Drive' *International Journal of Electrical Engineering & Technology* (IJEET) ISSN 0976 – 6553 Vol. 3, Issue 2, July – September (2012).

94. R.A. Jabbar Khan, A. Mohammed, M. Junaid, M. A. Masood and A. Iftkhar 'Lab VIEW based Electrical Machines Laboratory for Engineering Education' WSEAS Transactions on Advances in Engineering Education ISSN: 1790-1979 Issue 5, Vol. 7, May 2010.

95. M.H.N. Talib, Z. Ibrahim, N. Abdul Rahim and A.S. Abu Hasim 'Analysis on Speed Characteristics of Five Leg Inverter for different carrier based PWM Scheme' IEEE International Power Engineering and optimization Conference, Melaka, Malaysia: 6-7 June 2012.

96. Ramesh Reddy K., Neelashetty Kashappa, 'Performance of Voltage Source Multilevel Inverter-Fed Induction Motor Drive using Simulink', *ARPN Journal of Engineering and Applied Sciences*, Vol. 6, June 2011.

97. Matthew W. Dunnigan , Scott Wade and Barry W. Williams, 'Modeling and Simulation of Induction Machine Vector Control with Rotor Resistance Identification', IEEE transactions on power electronics, Vol. 12, No. 3 May 1997.

98. M.C. Trigg and C.V. Nayar, 'Matlab Simulink Modeling of a single-phase Voltage Controlled Voltage Source Inverter', Dept. of Electrical Engineering, Curtin University of Technology.

99. Slemon, G.R. 'Electrical Machines for Variable-Frequency', in Proceedings of the IEEE,Vol. 82, No. 8,pp.1123-1138, 1994.

100. Mohamayee Mohapatra and B.C. Hitti Babu, 'Fixed and sinusoidal-band hysteresis current controller for PWM voltage source inverter with LC filter', Member IEEE, 2010.

101. D.M. Deshpande and A. Ansari ,'Mathematical model of Asynchronous Machine in MATLAB Simulink", *International Journal of Engineering Science and Technology* Vol. 2(5), 2010.

102. M.A. EL-Barky and S.H. Arafah 'Simulation and Implementation of Three Phase Three Level Inverter' SICE July 25- 27, 2001.

103. A.M. Massoud, S.J. Finney and B.W. Williams 'Control techniques for multilevel voltage source inverters' IEEE 2003.

104. Jih-Sheng Lai, Tina Hua Liu and Siriroj Sirisukprasert 'Optimum harmonics reduction with a wide range of modulation indexes for multilevel converters' IEEE Trans Ind. Application Electronics, vol. 49, No. 4, August 2002.

105. G. Bhuvaneshwari and Nagaraju 'Multilevel inverters a comparative study' Vol. 51 No.2 march-April 2005

106. M. Depen brock, 'Pulse width control of a three-phase inverter with non sinusoidal phase voltage of a Three-phase PWM inverter', IEEE International Semiconductor Power Conversion Conference, Orlando, Florida, USA, 1977.

107. A.M. Hava, 'Carrier based PWM-VSI drives in the over modulation region', PhD Thesis, University of Wisconsin-Madison, 1998.

108. T.A. Lipo and D.G. Holmes, 'Pulse width modulation for power converters' principles and practices, IEE Press, Wiley Publications, New York, USA. 2000

109. J. Holtz, 'Pulse width modulation — a survey' IEEE Trans. on Industrial Electronics, Vol. 39, No. 5, Oct. 1992.

110. Kothari D.P. and I.J. Nagrath, *Theory and Problems of Basic Electrical Engineering*, Prentice-Hall of India, New Delhi, 1998.

111. M.P. Kazmier Kowski, R. Krishnan and F. Blaabjerg, '*Control in Power Electronics- Selected Problems*', Academic Press, California, USA. 2002.

112. R.J. Kerkman, B.J. Seibel, D.M. Brod, T. M. Rowan, and D. Branchgate, 'A simplified inverter model for on-line control and simulation', IEEE Trans. Ind. Applicant, Vol. 27, No. 3, pp. 567–573.1991.

113. T.J.E. Miller, Switched Reluctance Motors and Their Control, Magna Physics Publishing and Clarendon Press, Oxford, 1993.

114. R. Krishnan, *Switched Reluctance Motor Drives: Modelling, Simulation, Analysis, Design, and Applications*, CRC Press, 2001.

115. Lloyd, J., SR drive applications, in Tutorial Course on Switched Reluctance Motor Drives, organized by R. Krishnan, Conf. Rec. IEEE Ind. Appl. Soc. Ann. Mtg., Oct., 210–245, 1996.

116. Kothari D.P. and I.J. Nagrath, *Basic Electrical Engineering*, 3rd edn, TMH, New Delhi, 2010.

117. Krishnan, R. and A.S. Bharadwaj, A comparative study of various motor drive systems for aircraft applications, Conf. Rec. IEEE Ind. Appl. Soc. Ann. Mtg., Oct., 252–258, 1991.

118. Radun, A.V., High power density switched reluctance motor drive for aerospace Applications, IEEE Trans. on Ind. Appl., 113–119, 1992.

119. Ferreira, C.A., S.R. Jones, B.T. Drager, and W.S. Heglund, Design and implementation of a five horsepower, switched reluctance fuel-lube, pump motor drive for a gas turbine engine, Conf. Rec. IEEE Appl. Power Electronic Conf., 56–62, 1994.

120. Richardson, K.M., C. Pollock, and J.O. Fowler, Design and performance of a rotor position sensing system for a switched reluctance marine propulsion unit, Conf. Rec. IEEE Ind. Appl. Soc. Ann. Mtg.,Oct., 168–173, 1996.

121. Mauch, K. and F. Peabody, A compact thruster for industrial submarines, Conf. Rec. IEEE Ind. Appl. Soc. Ann. Mtg., Oct. 242–247, 1985.

122. Kothari D.P. and B.S. Umre, Lab Manual for Electric Machines, IK. Int., New Delhi, 2013.

123. Lee, B.S., H.K. Bae, P. Vijayraghavan, and R. Krishnan, Design of a linear switched reluctance machine, Conf. Rec. IEEE Ind. Appl. Soc. Ann. Mtg., Oct. 2267–2274, 1999.

124. Bae, H.K., B.S. Lee, P. Vijayraghavan, and R. Krishnan, A linear switched reluctance motor: converter and control, Conf. Rec. IEEE Ind. Appl. Soc. Ann. Mtg., Oct., 547–554, 1999.

125. Neha Thakur and Rakesh Singh Lodhi, Computer Simulation of SPWM-VSI for Minimizing the starting torque and current in Asynchronous Motor Drive, International Journal of Research (IJR) Vol-1, Issue-5, June 2014 ISSN 2348-6848.

126. Sourabh Mehto and Rakesh Singh Lodhi, Comparision and Analysis of Total Harmonic Distortion for IGBT and MOSFET based VS Inverter, Institute of Research in Engineering and Technology (Iret).

127. Payal Thakur and Rakesh Singh Lodhi, Review of Sensorless Vector Control of Induction Motor Based on Comparisons of Model Reference Adaptive System and Kalman Filter Speed Estimation Techniques International Journal of Computer Architecture and Mobility (ISSN 2319-9229) Vol. 1-Issue 6, April 2013.

128. Umashankar S, Anshuman Battacharjee, Anurag Chander Sharma, D. Vijaykumar and Kothari D P, "Mathematical Modeling of IGBT/Diode Inverter Fed Induction Motor Drive with its Switching Function using Decoupled Analytical Method," Paper presented in IEEE International conference on energy efficient technologies for sustainability (ICEETS'13), organized by St. Xaviers's Catholic College of Engineering, Nagercoil, Tamil Nadu, India on April 10-12, 2013.

129. LFC of an Interconnected Power System with Thyristor Controlled Phase Shifter in the Tie Line, (K P Singh Parmar, S Majhi and D P Kothari) *International Journal of Computer Applications,* ISSN 0975 – 8887 (online), Vol. 41, No. 9, pp 27-30, March 2012, Published by Foundation of Computer Science, New York, USA.

130. Umashankar. S., Ch. Bhanu Prasad, D. Vijayakumar and D.P. Kothari, "Reduction of Peak current of PFC Converter fed Induction Generator for Wind Power Generation excited by voltage source converter", publication in *International Journal of Applied Engineering Research* (IJAER), 2011.

131. G. Swaminathan, Umashankar S., Ramesh V., D. Vijayakumar and Kothari D.P., "On line condition monitoring and control for on load tap changer motor & drive system through Process bus", publication in 2nd International Conference on Advances in Energy Engineering (ICAEE2011), December 27-28, 2011, Bangkok , Thailand.

Index